Handbook of Laboratory Animal
Management and Welfare

Handbook of Laboratory Animal Management and Welfare

SARAH WOLFENSOHN BSc, MA, VetMB,
CertLAS, MRCVS
MAGGIE LLOYD MA, VetMB, MRCVS
University of Oxford
Veterinary Services

OXFORD UNIVERSITY PRESS
Oxford   New York   Tokyo
1994

Oxford University Press, Walton Street, Oxford OX2 6DP

Oxford   New York   Toronto
Delhi   Bombay   Calcutta   Madras   Karachi
Kuala Lumpur   Singapore   Hong Kong   Tokyo
Nairobi   Dar es Salaam   Cape Town
Melbourne   Auckland   Madrid

and associated companies in
Berlin   Ibadan

Oxford is a trade mark of Oxford University Press

Published in the United States
by Oxford University Press Inc., New York

A catalogue for this book is available from the British Library

Library of Congress Cataloging in Publication Data
Wolfensohn, Sarah.
Handbook of laboratory animal management and welfare / Sarah
Wolfensohn and Maggie Lloyd.
Includes bibliographical references (p.   ) and index.
1. Laboratory animals–Handbooks, manuals, etc.   2. Animal welfare–
Handbooks, manuals etc.   3. Animal experimentation–Handbooks,
manuals etc.   I. Lloyd, Maggie.
SF406.W64 1994   636.088′5–dc20   94-4697

ISBN 0 19 854833 8 (Hbk)
ISBN 0 19 854832 X (Pbk)

Typeset by The Electronic Book Factory, Fife, Scotland
Printed in Great Britain by Bookcraft (Bath) Ltd., Misomer Norton, Avon

This book is dedicated to improving the quality
and life of all those animals which are involved in
medical research.

Thanks are due to Mandy Dumighan for dedication to the word processor above and beyond the call of duty, and to Laurence Waters for his photographic skills.

CONTENTS

1 Introduction: Training and the personal licensee

GENERAL INFORMATION

The Home Office has made it clear that all applicants for personal licences which will enable them to carry out procedures on living animals, must receive formal training, sufficient to enable them to take primary responsibility for the animals. This training and education is to improve both animal welfare and the quality of science that is carried out.

Training can be achieved through lectures, tutorials, discussions, videos, and reading, but the importance of practical experience must be emphasized. The aim of this book is to supplement other training methods and bring useful information together in one volume, but the reader is encouraged to observe others and gain as much experience as possible.

New licensees should:

(1) have practical experience in handling and manipulating the commonly used laboratory species or any other species named in the licence application (see Chapter 9);

(2) be able to sex and age the animals to be used (see Chapter 6);

(3) understand the normal husbandry and nutritional requirements of the animals, together with their social, behavioural, and environmental needs (see Chapter 6);

(4) be able to recognize when the animal is deviating from normal behaviour and showing signs of ill health or distress, and know what action to take to prevent pain, or suffering (see Chapter 11);

(5) have a knowledge of the relevant legislation controlling the use of animals (see Chapter 2);

(6) understand the local rules in the institution relating to health and safety, security, supply, and disposal of animals (see Chapter 4);

(7) understand the importance of disease prevention in the animals and know why disease patterns are monitored (see Chapter 7);

(8) have a working knowledge of acceptable methods of anaesthesia and analgesia for the species used and procedures to be carried out (see Chapter 12);

(9) know how to dose and take samples from the species involved with due regard for volumes and sites and different methods for different species (see Chapter 10);

(10) some licensees will need more advanced instruction on methods of anaesthesia and analgesia for recovery surgery and should understand the principles of asepsis and surgical techniques. Suturing and surgical techniques will be relevant to some procedures (see Chapter 14 and 15), as will the importance of post-operative care (see Chapter 13).

The Home Office training requirements have been divided into modules which should be studied according to the type of work to be undertaken as shown in Table 1.1.

The chapters in this book correlate to the Home Office modules as follows:

Module 1. Chapters 1, 2, 3.

Module 2. Chapters 4, 5, 9, 11.

Module 3. Chapter 6, 7, 8, 10, 12 (part), 13 (part).

Module 4. Chapters 12, 13, 14, 15.

Module 5. is not dealt with, since this book is aimed at the **personal** licensee.

CONTENT OF HOME OFFICE MODULES:

Module 1. Legislation

1. Historical background.
2. Introduction to ethical aspects of the use of animals in experimental procedures.
3. The Animals (Scientific Procedures) Act 1986.
4. Other legislation relevant to the use of animals.

Module 2. Principles of animal care

1. Recognition of well-being, pain, suffering, and distress in relevant species.
2. Handling and restraint of relevant species.
3. Humane methods of killing.

Table 1.1
Home Office training requirements

Target Audience	Module 1	Module 2	Module 3	Module 4	Module 5
Those not applying for licence					
Personnel with administrative responsibility only	✓				
Non-licensed animal users: Those killing by a Schedule 1 method	✓	✓			
Licence applicants					
Undergraduates applying for limited personal licences to work under close supervision	✓	✓			
Personal licence applicants who will be performing minor non-surgical procedures; brief terminal procedures under anaesthesia	✓	✓	✓		
Personal licence applicants who will be performing major surgical procedures under terminal or recovery anaesthesia	✓	✓	✓	✓	
Project licence applicants	✓	✓	*Individual needs may vary*		✓

    4. Local procedures for security, administration, supply, and disposal of animals.

    5. Health and safety.

Module 3. Principles of animal use

    1. Biology and husbandry of relevant species.

    2. Common diseases in the relevant species.

    3. Health monitoring and disease prevention and control.

    4. Introduction to anaesthesia and analgesia.

    5. Conduct of minor procedures.

Module 4. Surgery

    1. Surgical anaesthesia and analgesia.

    2. Conduct of surgical procedures.

Module 5. Project management

    1. Further ethical considerations.

    2. Analysis of the scientific literature.

    3. Alternatives to using animals—the three Rs.

    4. Project design

    5. Project licence management.

    6. European and international legislation.

2 Legislation and ethical considerations

The primary piece of legislation which controls the use of animals in scientific experiments and with which licensees must be familiar is the Animals (Scientific Procedures) Act 1986.

HISTORICAL BACKGROUND TO THE ANIMALS (SCIENTIFIC PROCEDURES) ACT 1986

Late in the 17th Century, laws were passed in the United Kingdom against cruelty to animals. In 1822, Martin's Act (after its sponsor Richard Martin) was passed which primarily protected cattle and horses and was amended in 1835 to protect all domestic animals. The first prosecution against use of animals in experiments was in 1874, when the RSPCA brought an unsuccessful case against the French physiologist, Eugene Magnan.

In 1831, British physiologists attempted self-regulation with the publication of Marshall Hall's five principles (see Rupke 1987) and a moral code of practice for work with experimental animals was drawn up by the Committee of the British Association for the Advancement of Science, which published guidelines in 1871.

The Cruelty to Animals Act, which related to experiments expected to cause pain to living animals, was passed in 1876. The main differences from the present legislation were that 'experiment' was not clearly defined and that initially the main concern was only over surgical procedures. However, there was increasing concern over the wide variety of types of non-surgical work being performed. Pain was not clearly defined in the 1876 Act and there was little control over the purposes for which techniques could be used once permission had been given. There was no explicit justification required in order to limit the suffering caused in pursuit of relatively trivial purposes. There was no control over the experiment's scientific design, the number of animals or the species used, or the competence of licensees.

The law only applied while the animal was under experiment,

there was little control over the animals' care and welfare outside this time, and no control over the breeding and supply of experimental animals.

In 1926, Charles Hume founded the University of London Animal Welfare Society (now the Universities Federation for Animal Welfare—UFAW) to try to get people to think rationally about their attitude to animals. He published the first edition of the UFAW *Handbook on the care and management of laboratory animals* in 1947 (now in its Sixth edition). He also commissioned Russell and Burch to write *The principles of humane experimental technique* in 1959, which brought out the concept of the three Rs: replacement, reduction, and refinement.

Despite its limitations, the 1876 Act stood for 110 years, largely due to interpretation by the Home Office, which had to take account of major scientific developments. By the 1960s a strong case for updating the legislation was made and the Littlewood Committee reported that the provisions of the Act 'had not matched up with modern scientific and technological requirements'. Certain administrative changes were made including strengthening the Inspectorate and setting up an independent Advisory Committee. Pressure for change continued into the 1970s and in 1976, coinciding with 'Animal Welfare Year' and the Centenary of the 1876 Act, the Committee for Reform of Animal Experimentation (CRAE) was formed. Lord Halsbury brought a Private Members' Bill before Parliament in 1979 and in 1980 the Home Office Advisory Committee was asked to make recommendations for reform. In 1983, CRAE, the British Veterinary Association (BVA) and the Fund for Replacement of Animals in Medical Experimentation (FRAME) joined forces to negotiate the White Papers and the Animals (Scientific Procedures) Act was passed in 1986.

SUMMARY OF THE MAIN PROVISIONS OF THE ANIMALS (SCIENTIFIC PROCEDURES) ACT 1986

This piece of legislation controls PROCEDURES that may cause PAIN, SUFFERING, DISTRESS OR LASTING HARM on all LIVING VERTEBRATES other than man, including immature forms from specified stages in their development. Octopus vulgaris from the time of hatching are also protected by the Act.

TWO kinds of licence are required for all work regulated by the Act:

  1. A PERSONAL LICENCE (PIL) for individual workers specifying procedures, and types of animals they may use according to competence, training, qualifications, experience, and general suitability. *The personal licensee bears primary*

responsibility for the care of animals on which they have carried out scientific procedures.

2. A project licence (PPL) covering each specific programme of work which specifies:
   (a) the purpose and scientific justification for the work;
   (b) a full description of the procedures involved (experimental protocol);
   (c) an estimate of the number of animals of each species which will be required;
   (d) an assessment of potential severity of each procedure and of the project as a whole;
   (e) qualifications and experience of the project licence holder and deputy to assess suitability.

Each project licence application requires specific scrutiny by the Home Office Inspectorate to weigh POTENTIAL BENEFIT VERSUS ANY PAIN, SUFFERING OR HARM (known as ADVERSE EFFECTS) that may be caused to the animals used. For each application it has to be stated, for example, whether the project will be for control of disease (medical benefit), for physiological studies (scientific benefit), for environmental protection (environmental benefit) or for education and training (educational benefit).

There are controls on pain and suffering as the project is graded into one of three SEVERITY BANDS: mild, moderate or substantial, and *the personal licensee must observe the animal to ensure the allocated severity band is not exceeded* (see Chapter 11—recognition of pain and stress). Where all the procedures in the project licence are performed under terminal anaesthesia or on decerebrate animals, then the overall severity is unclassified. Anaesthetics, analgesics, and other methods of preventing or minimizing pain must be used whenever possible (see Chapters 12 and 13). The three severity bands are not discrete entities but form a spectrum across a wide range of adverse effects. It is necessary to consider not only the immediate effects on the animal, but the longer-term effects as well, in order to gauge the overall effects of the procedure. For example, establishing a ruminal fistula has a high adverse effect initially but low adverse effect long term; whereas the effect of injecting a carcinogenic drug may be low initially but very severe in the longer term (see Figure 2.1).

In order to assess the severity of the procedure it is necessary to take into account this overall effect and the likely incidence of its occurrence. For instance, by limiting the occurrence of severe adverse effects to only 1 per cent of the animals involved, and humanely killing any animal which begins to exhibit more than mild adverse effects, it may be possible to place the procedure into the mild severity band. This introduces the concept of *the humane end* point which refines the experimental technique in order to reduce the overall suffering. A predetermined level of adverse effects is set down and if an animal starts to exceed this, it is killed (see Chapter 11 for further details on distress scoring).

Using these parameters, judgement is made about the level of potential benefit versus the level of potential suffering (Bateson 1986) and, if satisfied, the Inspectorate then recommends authorization of the project by the Secretary of State.

Examples of severity of common procedures are:

> *Mild:*
> – small or infrequent blood sampling
> – superficial biopsy
> – procedures in which the animal will be terminated before it shows more than minor changes from its normal behaviour (use of the *humane end point*)
> – minor surgical procedures under anaesthesia

Assessing the potential effect

Overall effects of procedure

Immediate adverse effects of procedure			
high	medium	high	high
medium	low	medium	high
low	low	medium	high
	low	medium	high

Long-term effects of procedure

Assessing the severity

Severity band

Overall effect of procedure			
high	mild	moderate	substantial
medium	mild	moderate	moderate
low	mild	mild	mild
	<5%	5-25%	>25%

Incidence of adverse effects of procedure

2.1 Assessing the potential effect and severity on the animal.

Moderate:
– surgical procedures where post-operative care and analgesia are reliably provided.
– toxicity tests with defined *humane end points* (as opposed to lethality as the end point).

Substantial:
– major surgery causing post-operative suffering.
– toxicity studies with significant morbidity or death as an end point.
– any procedure which results in significant deviation from the animals' normal state of health.

If the severity limit of the procedure is exceeded the Home Office Inspector must be informed.

Project licences will not be issued unless the applicant has considered the use of ALTERNATIVE methods not involving live animals and none has been found to be feasible.

Establishments for both experimental work and those involved in breeding and supply of Schedule 2 laboratory animals (see Figure 2.2) are controlled by a **Certificate of Designation** issued to safeguard the standards of care and welfare for the animals kept there. This certificate indicates which rooms may be used for what purposes, and are classified according to:

LTH Long-term holding

STH Short-term holding (less than 48 hours)

NSEP Non-sterile experimental procedure

SEP Sterile experimental procedure

Mouse
Rat
Guinea-pig
Hamster
Rabbit
Dog
Cat
Primate
Common Quail (Coturnix coturnix)

2.2 Schedule 2: Animals to be obtained only from designated breeding or supplying establishments.

SF   Service facility

SA   Small animal

PRI   Primate

CAT   Cat

DOG   Dog

LA   Large animal

The Certificate of Designation is issued to the **Certificate Holder**, usually someone in a senior management position, who is ultimately responsible for ensuring that work carried out at the Designated Premises complies with the requirements of the Act. The Certificate also identifies the **Named Veterinary Surgeon** and the **Person in Day to Day Care** by name. The standard conditions under which the Certificate of Designation is issued can be found in Appendix II (Scientific Procedure Establishments) and Appendix III (Breeding and Supply Establishments) of the Home Office Guidance Notes.

You are obliged to read the Guidance on the Operation of the Animals (Scientific Procedures) Act 1986 (HMSO (1990:). HC 182, ISBN 010 2182906) **before applying for a licence** (referred to as 'the Guidance' in Home Office notes).

STATISTICS OF ANIMAL USAGE (at August 1993)

People working with experimental animals should be aware of some of the facts concerning their use:

> There are 358 designated Scientific Procedure establishments in the United Kingdom, with about 4400 project licences and 17 000 personal licensees covering the work that is done.

> There are 21 Home Office Inspectors who made 3299 visits to establishments in 1992.

> A total of 2.94 million animals were used in procedures in 1992. Nearly 80 per cent of procedures were carried out on rats and mice. Other animals used included other rodents (5 per cent), birds (8 per cent), fish (5 per cent), rabbits (3 per cent), dogs (0.31 per cent), cats (0.13 per cent), and primates (0.17 per cent).

> 68 per cent of all procedures were carried out for medical and veterinary research. 14 per cent were concerned with the production and development of biological materials used in treatment and research. 6 percent of studies were for breeding transgenic animals and animals with genetic defects. 9 percent were for non-medical safety testing of industrial, agricultural, and other products. Testing

cosmetics accounted for 0.07 per cent of all animal procedures (about 2200 procedures).

THE ETHICS OF USING ANIMALS IN EXPERIMENTS

All use of animals in scientific research for human benefit creates a dilemma—the justification for using the animal depends on it being different from the human, while the validity of the results obtained depends on the similarity of the animals and their responses to those of the human.

As well as being controlled by law, the use of animals for scientific purposes also depends on individual standards. When considering the experiment and the project licence application, a declaration has to be signed on the form that the purpose of the project could not be achieved without involving the use of animals.

This section is not intended to provide a comprehensive guide to the ethics of using animals, and its brevity should not be taken to indicate that such considerations should be treated lightly. This book is intended as a practical handbook and the reader is urged to consult the list of further reading and to examine the texts for a more detailed consideration of the ethics of animal usage.

In the meantime, it is important to consider the following questions before embarking on the experiment:

1. *Why are you doing it?*
   Can you justify what you are doing and does the potential benefit derived from the result outweigh the cost to the animal?

2. *Does what you propose to do raise ethical issues?*
   Is the planned experiment going to cause the animal any pain or any other form of suffering, such as isolation or confinement?

3. *Is replacement possible?*
   Is it possible to achieve the same benefit by carrying out an *in vitro* experiment or using a mathematical model or a human subject or a different animal species lower in the phylogenetic scale?

4. *Is reduction possible?*
   Can you use less animals by better experimental design, or collaborating with colleagues so that as many tissues as possible from each animal are used. Are 'spare' breeding animals and excess young produced used? Can you get Home Office approval for re-use of some animals?

5. *Is refinement possible?*
   Can you alter your experimental method in some way to decrease the animal's potential suffering, for example, by

an alteration in surgical technique (e.g. using laparoscopy instead of a laparotomy), alterations in housing or bedding or altering the drug dosing regime for a reduced volume or reduced frequency?

At all times throughout the experiment consider what you can do to decrease the potential for suffering inflicted on the animals.

OTHER RELEVANT LEGISLATION

There are many other pieces of legislation, regulations, and guidelines which may be relevant to certain areas of laboratory animal use.

The British laws relevant to using laboratory animals are ones which cover:

1. Protection of domestic/captive animals
   - Protection of Animals Act 1911-1964
   - Veterinary Surgeons Act 1966

2. Control of animal experiments
   - Animals (Scientific Procedures) Act 1986

3. Control of animal disease/zoonoses
   - Animal Health Act 1981

4. Importation and transport of animals
   - Animal Health Act 1981: Rabies (Importation of Dogs, Cats and Other Mammals) Order 1974
   - Rabies Control Order 1974
   - Transit of Animals (General) Order 1973
   - Transit of Animals (Amendment) Order 1988
   - Movement of Animals (Records) Order 1960
   - Movement and Sale of Pigs Order 1975

5. Control of dangerous wild animals
   - Dangerous Wild Animals Act 1976

6. Human safety—Home Office
   - Health and Safety At Work Act 1974
   - Control of Substances Hazardous to Health Regulations 1989

There is an absolute requirement under the HSW Act and COSHH to assess and control the hazards and risks in the work place.

7. Conservation of endangered species and protection of wild animals
   - Convention on International Trade in Endangered Species (CITES)
   - Wildlife and Countryside Act 1981
   - Badgers Act 1973

8. Control of dangerous drugs and firearms
   - Poisons Act 1972
   - Misuse of Drugs Act 1971
   - Medicines Act 1968
   - Firearms Regulations

There are also regulations covering transport of animals (International Air Transport Association IATA) and good laboratory practice (GLP).

This list is not exhaustive, so for further information, consult the list of further reading.

RECORD KEEPING

Under the Animals (Scientific Procedures) Act 1986, it is a requirement that certain records are maintained by the licence holders. Records should be kept for a period of **5** years following the death of an animal or its release from the establishment, and should be kept for **each protected animal** either individually or in batches.

For dogs, cats, horses, primates, cattle and other farm animals, and adult birds, each animal must be readily identifiable by an approved method of permanent marking. Such methods include:

> tattoo
>
> collar
>
> microchip implant
>
> freeze branding
>
> ear tagging
>
> leg ringing, as appropriate for the species

Each cage or area holding animals not undergoing a regulated procedure should bear a label identifying:

> cage/area identification
>
> identification of animals held (individual or batch)
>
> date entry made

The following information should be recorded for normal animals, pretreated animals or those with harmful genetic defects:

> source
>
> species
>
> breed or strain
>
> identification (batch or individual)
>
> arrival date

age on arrival

sex

if female, whether or not pregnant

dates in or out of quarantine or isolation, if applicable

microbiological status

pretreatment

harmful genetic defects

Project Licence (PPL) number

date and method of disposal/re-use/release

Health records must be maintained in consultation with the named veterinary surgeon and must be available to the day-to-day care person.

Project licence records should include:

PPL number, name of holder, and deputy

Names of personal licensees involved

Details of procedure:

– species

– number of each species

– sex and age at commencement of procedure

– identification of animal (or batch)

– date of start of procedure

– any unexpected morbidity or mortality

– re-use within the project

– date of end of procedure

Fate of animals at end of procedure:

A: released to the wild

B: released to private care

C: released for slaughter

D: killed within establishment

E: Re-use, identify PPL to which transferred

Personal licence holders are responsible for ensuring that all cages or enclosures are adequately labelled. The label must identify:

– the project

– the personal licensee

– the procedures

A coded system may be used.

Full details of record keeping requirements can be found in Appendix V of the Home Office Guidance notes.

An example of a Project Licence Record Chart design is in Figure 2.3. Use of this format enables the information needed for the annual returns to be easily retrieved.

NOTES TO ASSIST IN COMPLETION OF THE PERSONAL LICENCE APPLICATION FORM

You may not begin a procedure coming within the scope of the Act until you have obtained a personal licence from the Secretary of State. Your personal licence may only be used in conjunction with a valid Project Licence. Scientific procedures other than those which cause pain, suffering, distress or lasting harm to protected animals do not require the authority of a licence.

No one under the age of 18 may hold a licence. You should list your qualification(s). If you are not a graduate, give full details of your educational qualifications: applicants should normally possess the equivalent of at least five GCSEs, or have received appropriate formal vocational training.

PART I
The Supervisory Condition (Questions 10 and 11)

Many licensees will be subject to a condition of supervision by a personal licence holder who has held a licence for at least 1 year and who works closely with them. A condition of supervision will apply to you if you have not previously held a licence, and will normally apply if you come from outside the United Kingdom and have not lived in this country for the past 5 years. A condition of supervision may also be imposed or reimposed if you subsequently seek authority for techniques other than those for which you originally applied. The majority of new licensees will be placed under supervision in order to ensure the attainment of competence, but in the case of experienced scientists from abroad, the purpose of supervision is to enable them to be given guidance about the requirements of the Act.

A supervision condition will normally remain in force for 1 year unless it is lifted before that time on the supervisor's application. The Secretary of State will lift the condition when he/she is satisfied that the licensee has attained a sufficient level of competence to perform without supervision the techniques for which he/she has authority or, in the case of experienced scientists from abroad, when they have become familiar with working in accordance with the requirements of the Act. While the supervision condition remains in force, the level of supervision is a matter for the supervisor, in consultation with the inspector if necessary. At first, the supervisor may need to observe and advise a licensee throughout the performance of a technique or procedure. Later on, when the supervisor is satisfied with the licensee's technical proficiency he/she may reduce the level of supervision to monitoring and discussing his work. Finally, he/she may ask the Secretary of State to lift the supervision condition altogether. Undergraduate licensees remain under supervision until they have completed their degree course.

Department:

Project licence no.:

Page no.:

Animal I.D. no.	Personal licence holder	Species and strain	No. in exp.	Sex	Age or wt.	Date of procedure	Procedure(s) (Brief details of anaesthetic, etc.)	Date procedure end	Re-use? (If allowed)	Date of disposal	Disposal details	Comments (Clinical and veterinary observations and notes)

2.3 Project licence record.

The purpose of supervision is to ensure the competence and reliability of new personal licensees and the protection of the animals they use. Regardless of whether or not they are under a supervision condition, all personal licensees who are not also holders of a project licence should obtain guidance from the project licence holder about the way in which techniques and procedures forming part of the project are to be carried out. The project licence holder has a general responsibility to ensure that personal licensees working on his project carry out their work properly and humanely and keep within the terms of the project licence.

Undergraduate student licensees should nominate supervisors who hold a personal licence and will be named in the conditions attached to the licence. Other licensees may normally be supervised by the heads of the departments or equivalent in which they work, or senior licensees acting as their deputies. The licence condition in these circumstances will not normally name the supervisor, but a supervised licensee should at all times know who is acting as his/her supervisor.

The imposition of a supervision condition will not lessen the individual responsibility of the personal licensee to comply with the provisions of the Act and the terms and conditions of the licence.

Question 12
A licence is normally granted for an indefinite period but will be subject to review at periods not exceeding 5 years. Licences for undergraduate students are subject to annual review.

Question 13
Applicants are granted authority for techniques appropriate to the type of work they intend to perform. If you are not simultaneously submitting a project licence application, you should, if possible, specify the projects on which you will be working. Even if the project on which you intend to work has not yet been granted a project licence, give details of the proposed project. Project licence holders have the responsibility of holding a list of personal licensees working on their projects.

Question 14
The Act requires procedures to be performed at Designated Places unless exceptionally, the type of work requires them to be performed elsewhere. Permission will have to be obtained under both the personal and project licences. You should therefore list *all* places at which you intend to perform procedures, drawing attention if necessary, to the possibility that you may need to work at a non-designated place. If you propose to carry out work at an undesignated location, you should describe the location as accurately as possible. If after your licence has been issued you wish to work at a place not

specified on it, the licence will require amendment. The licence must cover all of the techniques and species you need to use as, once it is granted, it will define the entire extent of your personal authority.

Question 15
Reference to the details on the project licence will aid in describing the techniques in this section.

(a) A technique is a technical act or omission, e.g. dosing, bleeding, laparotomy, withholding food or water. Number and list each technique with a succinct description. For the administration of substances (including anaesthetics) or removal of body fluids, each method and route should be specified. The administration of an anaesthetic is a regulated procedure in its own right. Associated surgical techniques such as tracheostomy, should be listed separately. Some examples are:

	Technique	**Animal**	**Anaesthesia**
(i)	Administration of substances by admixture with food or water, by gavage or injection by the following routes: subcutaneous, intramuscular, intravenous, etc. (specify)	Mouse Rat Rabbit	AA

If the same techniques are to be carried out sometimes with, sometimes without, anaesthesia, the anaesthetic code might be 'AA/AB' or 'AA/AB/AC', indicated 'GENERAL' or 'LOCAL' (see section (c) 4).

(ii)	Withdrawal of blood by superficial venepuncture or venesection or via catheters implanted under separate authority	Rat	AA/AB (General)

or

Withdrawal of body fluids by superficial venepuncture or by peritoneal aspiration.

(iii)	Induction and maintenance of general anaesthesia by inhalation, intraperitoneal injection, intravenous injection, etc. (specify route)	Rat Rabbit	AA/AB/AC (General)

or

Administration of a general
anaesthetic by injection and/or
inhalation.

More complex techniques and those requiring a surgical approach such as the administration of substances enterally, parenterally (administration into the brain or eye must be specified) or by external application (application to the eye must be specified) and withdrawal of body fluids (withdrawal from the heart, orbital structures, or brain must be specified), must be noted separately. Killing for an experimental purpose at a designated establishment is a regulated procedure unless an appropriate method from Schedule 1 of the Act is used. Techniques involved in such killing must therefore be listed.

If you intend to use any unusual techniques or novel or complex surgery, describe them in detail with additional diagrams if they may be helpful. Such techniques may require separate justification in the project licence.

(b)   The kinds of animal should be listed as follows:
MAMMALS
mouse
rat
guinea-pig
hamster
gerbil
other rodents (indicate species)
rabbit
cat
dogs
– beagle
– greyhound
– other types
ferret
other carnivores (indicate species)
horse, donkey, and crossbreds (*Equidae*)
pig
goat
sheep
cattle
other ungulates (indicate species)
primates
– prosimians
– New World monkeys:
   marmoset, tamarin, squirrel monkey,
   owl monkey, spider monkey
   other New World monkeys

– Old World monkeys:
  macaque (rhesus, cynomolgus)
  baboon
  other Old World monkeys
– apes:
  gibbon
  great ape
other mammals (indicate species)

BIRDS (indicate genus)
REPTILES (indicate genus)
AMPHIBIANS (indicate genus)
FISH (indicate genus or other convenient grouping)

Embryonic, larval, and fetal forms are protected by the Act. In the case of mammals, birds, and reptiles, from halfway through the gestation or incubation period for the species; in the case of fish and amphibians, from the time at which they become capable of independent feeding. Procedures which begin before these points but continue, or have consequences continuing beyond them, require the authority of a licence. State 'embryonic', 'fetal' or 'larval' if you are using these forms.

(c) 1. Certain techniques may involve the use of anaesthesia in some kinds of animal but not in others, make this clear.
    2. A personal licence is required to anaesthetize or decerebrate an animal for subsequent scientific use.
    3. For all warm- and cold-blooded vertebrates, full anaesthesia is required for decerebration or other transections of the neuraxis (with or without destruction of rostral tissues)—but an exemption from this requirement may be considered for certain cold-blooded vertebrates for applicants showing sufficient competence and experience.
    4. Use these codes to describe the anaesthetic status and use of a neuromuscular blocking agent.
       AA   No anaesthesia when the technique is applied.
       AB   Anaesthesia when the technique is applied but with subsequent recovery.
       AC   Anaesthesia when technique is applied without subsequent recovery.
       AD   Use of neuromuscular relaxants.
    5. Distinguish between local, general, and regional anaesthesia.
    6. Indicate if general anaesthesia is required for the technique, but will be provided by another licensee.
    7. Enter as many of the above codes for each technique as are appropriate.

Question 16

The purpose of this question is to identify applicants who intend to delegate tasks not requiring technical knowledge to an assistant who is not a personal licensee, as you may not delegate subordinate duties unless your licence conditions authorize you to do so. See below for examples of procedures which may be delegated in this way.

PART II—should be completed by the sponsor.
Question 19-22

A sponsor for the applicant will normally be a senior scientist in the department or section in which the applicant for a personal licence will carry out the work and will always himself hold a personal licence.

All applicants are required to familiarize themselves with their responsibilities as licensees as much as possible in advance of applying for a licence. The duty to comply with the law is theirs, not the sponsor's. The Secretary of State merely seeks the sponsor's assurance, where necessary, that the applicant is competent, of suitable character and understands his duty under the law. Note that an applicant's proficiency may be acquired and demonstrated on dead animals but not on live ones. Personal licences are issued subject to Standard Conditions, which may be found in Appendix VI of the Home Office Guidance notes, and should be read carefully.

Examples of the kinds of non-technical procedures which may, with permission, be delegated to non-licensed assistants (see Appendix VII of the Home Office Guidance notes)

In certain circumstances, it may be possible for personal licensees to delegate the conduct of certain regulated procedures to non-licensees. The following list contains some examples. Wherever it refers to tasks 'previously' carried out, those tasks will have been specified by a suitably qualified personal licensee, *who must be within reach for assistance or advice if required.*

1. The filling of food hoppers and water bottles with previously mixed diets or liquids of altered constitution or to which test substances have been previously added.
2. The placing of animals in some previously set-up altered environments, e.g. inhalation chambers, pressure chambers, aquatic environments.
3. Pressing the exposure button to deliver previously determined doses of irradiation to an animal.
4. Pairing/grouping associated with the breeding of animals with harmful genetic defects.
5. Withdrawal of contents from an established ruminal fistula.

6. Operating automated machinery which carries out inoculation of eggs.

7. Placement of animals in restraining devices, as defined by the project licence.

8. Withdrawal of food and/or water, as defined by the project licence.

9. Placement of avian eggs into previously set chillers at the termination of a procedure.

Tasks which may be undertaken by assistants only in the *presence* of a suitably authorized personal licensee:

10. In animals rendered insentient by decerebration or general anaesthesia which is to persist until death, and through an established catheter, administration of substances as defined by the project licence or removal of body fluids.

11. In animals rendered insentient by decerebration or general anaesthesia which is to persist until death, the administration of electric stimuli through electrodes implanted by a personal licensee.

NOTES TO ASSIST WITH UNDERSTANDING THE PROJECT LICENCE APPLICATION

Although this handbook is aimed at the personal licensee, all work has to be authorized by a project licence. The following notes are to guide the personal licensee in understanding the scope of the project licence application.

The purpose of the application form is to provide information about the proposed use of regulated procedures so that the Secretary of State can ensure that the use of animals is appropriate, justified, and that the likely adverse effects on the animals concerned can be weighed against the benefit likely to accrue as the result of the programme of work, as is required by the Act.

It may be of benefit to consult the named veterinary surgeon and day-to-day care person in drawing up the project licence application, for advice on control of adverse effects of the procedures and on refinements of techniques.

Section 10

In addition to providing details of status, qualifications and experience, the applicant should bear in mind the following elements relevant to overseeing work on laboratory animals:

(a) ethical and legal considerations relevant to laboratory animal use;

(b) knowledge and skills required to design, conduct, and analyse scientific experiments using living animals;

(c) knowledge of alternative and refined methods available to replace procedures on living animals.

Section 11
Most project licences will require at least one deputy project licence holder, who will normally hold, or have held, a personal licence.

Section 12
Where it is necessary to conduct work at more than one designated establishment, either the licence holder or one of the nominated deputies should normally be based at each establishment and be in a position to direct work authorized under the project licence, and this section should be signed by the deputy project licence holder at each additional establishment.

Section 14
Under section 5(3) of the Act, the categories of permissible purposes are as follows:

(a) the prevention or diagnosis or treatment of disease, ill-health or abnormality or their effects in man, animals or plants;

(b) the assessment, detection, regulation or modification of physiological conditions in man, animals or plants;

(c) the protection of the natural environment in the interests of the health or welfare of man or animals;

(d) the advancement of knowledge in biological or behavioural sciences;

(e) education or training, other than in primary or secondary schools;

(f) forensic enquiries;

(g) the breeding of animals for experimental or other scientific use (i.e. where the breeding is a regulated procedure).

Work may fall into more than one category.

Section 16
A project licence is usually valid for a maximum of 5 years. They cannot be extended to last longer than 5 years and authority to continue work after the expiry of 5 years from the original date of issue will require a new licence, application for which must be made in sufficient time to allow due consideration, including possible referral to an assessor and/or the Animal Procedures Committee.

Section 17

Background: This should give a concise account of the present state of relevant knowledge, with key supporting references. Any previous work in the proposed field of study by the applicant or members of the group should be indicated, also with supporting references.

Objectives: This section should define the problems it is hoped to solve during the lifetime of the project in specific rather than general terms. Longer-term aims may be described separately.

Potential benefits: assuming that the objectives are met successfully, what short-and long-term benefits are likely to follow?

It should be clear from the text of section 17 that the proposed work is original and well justified by potential benefits.

Section 18

In describing the plan of work, it must be shown how the individual procedures will be used to meet the scientific objectives of the project. Fuller details of the procedures are required in section 19 and need not be set out in section 18. There should be details here of any control groups and statistical treatment of results. This section should establish that the proposed work is well designed scientifically and is likely to meet the objectives set out in section 17.

This section should also give details of how individual procedures inter-relate in the sequencing or staging of the project. Flow diagrams may be helpful.

For the purpose of this section and section 19, a procedure is defined as one or more techniques directed towards a common purpose. Thus the procedure of raising polyclonal antisera includes the techniques used to immunize and sample, all of which should be set out in detail in section 19b.

In all cases it must be shown that replacement of animals with non-sentient alternatives is not appropriate for any part of the project, and that full consideration has been given to reducing the number of animals used and to refining the procedures to minimize suffering. This consideration forms parts of the declaration in section 21 of the application.

Section 19

This section is in two parts: section 19a is an **index of procedures**, setting out the animals to be used and the severity of the procedures. The **protocol sheets** (19b) provide full details of these procedures and their constituent techniques; any adverse effects; and methods for controlling severity.

The application should, as far as practicable, be asking only for those procedures that are likely to be needed and can be defined in detail. Amendments to the project and procedures may be applied for later in the light of progress and experience with the project.

Section 19a: Index of procedures

1. Each procedure has a reference number.
2. Each procedure has a very short title.
3. There is one severity limit for each procedure. The severity limit should be either mild, moderate, substantial or, in the case of procedures carried out on decerebrate animals or wholly under terminal anaesthesia, unclassified.

As far as each procedure is concerned, the limit should be applied for which covers the maximum degree of severity likely to be seen in the animals and which is consistent with the objectives. It is not necessary to make allowance in the assessment for the unexpected. In complex procedures combining terminal and non-terminal techniques, the severity limit should relate only to the non-terminal techniques. Suggestions for the severity assessment of individual procedures are given in the Home Office Guidance notes, paragraphs 4.4 to 4.13. The severity limit in column 3 of section 19(a) should take into account details of the procedure itself, the nature of any likely adverse effects and the action to be taken to minimize these effects. Adherence to this limit and observance of any other controls described in section 19b are required by the standard licence condition governing severity.

4. The kind of animal to be used must be specified, also the animal's stages of development, and an estimate of the number to be used each year. Embryonic, larval, and fetal forms are protected by the Act; in the case of mammals, birds, and reptiles, from halfway through the gestation or incubation period for the species; in the case of fish and amphibians, from the time at which they become capable of independent feeding. Procedures which begin before these points but continue, or have consequences, beyond them, require the authority of a licence.

Section 19b: Protocol sheet

The purpose of this section is to give sufficient detail of what is to happen to each animal (or group of animals treated similarly) to allow a realistic assessment of the likely severity of the protocol's effects. It follows that the total use of the animal from the time of removal from stock to the time it is killed or otherwise discharged from control should be contained within one protocol description if at all possible.

There should be a separate protocol sheet (both pages) for each procedure listed in the index comprising 19a.

Each protocol sheet should be in a form such that it can be handed to a personal licensee as a working document indicating how the procedure should be carried out; what adverse effects are likely; and what must be done to remain within the severity limit authorized in section 19a.

Section 19b(v)

To give a complete picture it may be necessary to include details of some actions which are not themselves regulated. For techniques comprising regulated procedures, any details which might be relevant to an assessment of severity should be included. For example:

(a) brief description of all techniques indicating, by use of the anaesthetic codes listed below, whether with or without anaesthesia, and whether it is general, regional or local, and the route(s) of administration;

(b) classes of agents or compounds to be administered, route(s), frequency;

(c) sampling techniques: routes, volumes, frequency;

(d) catheters, cannulae: sites, temporary or indwelling, unilateral or bilateral, total number;

(e) CNS lesions: sites, unilateral or bilateral, how created, total number;

(f) nerve section: specify nerve(s), unilateral or bilateral;

(g) exposure to infection: agent(s), route(s);

(h) techniques which are to be carried out under terminal anaesthesia must be clearly identified by using the code AC;

(i) any repetition or combination of techniques on individual animals must be clearly stated (see also note 19b vii below);

(j) duration of procedures should be stated where relevant.

The above list is a guide only and is not intended to be exhaustive.

Anaesthetic codes

AA No anaesthesia during any part of the technique.

AB Anaesthesia at some time during or throughout the technique and with recovery of sensation. The distinction between general, regional, and local anaesthesia can be made by the additional letters G, R or L where appropriate.

AC General anaesthesia during the entire technique without recovery of consciousness.

AD   Any use of a neuromuscular blocking agent—this code must not be used without one of the other codes above.

Section 19b(vi)

In this section, all expected adverse effects should be listed together with methods of control, which may involve use of analgesics or the use of a defined humane end point when the animal is killed to prevent potential distress.

For further advice see Chapters 11 and 13 and discuss the proposal with the named veterinary surgeon.

Section 19b(vii)

Section 14 of the Act relates to the re-use of protected animals. A number of detailed considerations apply to animals which have been given a general anaesthetic for the first use and these are set out in paragraphs 4.21 to 4.29 of the Home Office Guidance notes.

Section 20

If all the individual procedures are under terminal anaesthesia or on decerebrate animals then the overall severity of the project is unclassified. Otherwise the severity of the project should be assessed as **mild**, **moderate** or **substantial**.

As far as the project is concerned, the assessment of overall severity should take account of the severity likely to be reached in all the individual procedures; the duration of that severity; and the number of animals affected. Overall severity thus has qualitative and quantitative aspects and does not simply equate to the severity of the single most severe procedure.

Section 22

All applications must be countersigned in section 22 by the Certificate Holder for the designated establishment or by a senior member of the staff of the establishment authorized by the certificate holder to carry out this function.

Assessment of the application

The Secretary of State has power under section 9 of the Act to refer the application to an external assessor and/or to the Animal Procedures Committee for advice. See paragraphs 4.30 to 4.31 of the Home Office Guidance notes.

You may not start any regulated procedure until a project licence covering the work has been granted by the Secretary of State.

Project licences are issued subject to standard conditions which may be found in Appendix IV of the Home Office Guidance notes.

3 Roles and responsibilities of 'named' persons

Many people are involved in the care of laboratory animals. The diagrams in each of the following sections show that communication between these groups of people is vital if everyone is to carry out their responsibilities efficiently. Individual institutions will vary in how their line management is organized but the various groups must communicate effectively. Remember, the animals cannot speak for themselves.

1. **The certificate holder**, i.e. the holder of the Home Office Certificate of Designation, has overall legal responsibility for all the animal facilities and for all the procedures carried out in an institute. The certificate holder is responsible for ensuring that the buildings in which experimental or breeding animals are kept are maintained to a satisfactory standard and that there are adequate numbers of suitably trained staff to care for the animals. It is the certificate holder who has to ensure that all procedures carried out in the establishment are authorized by project and personal licences, and that licensees are properly trained.

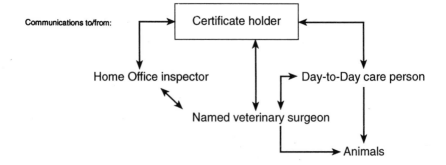

2. **The project licence holder** directs and supervises a particular work project. This person must make sure that the work is done in such a way as to keep within the conditions of the licence, and that only the species and number of animals listed on the licence are used. He or she must ensure that any personal licensees carrying out procedures within the project are appropriately licensed, trained, and supervised (if this is stipulated) to do so, and are aware of the severity conditions of the licence and that full records are kept (see Record keeping, Chapter 2).

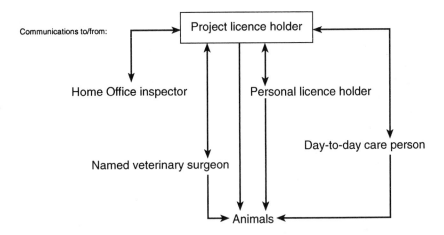

3. **The personal licensee** bears primary responsibility for the care of the animals on which he or she has carried out procedures. This person must ensure that the cages or pens in which the animals are kept are adequately labelled (see Record keeping, Chapter 2) and must be familiar with the severity limit and constraints on adverse effects in the protocol sheets of the project licence.

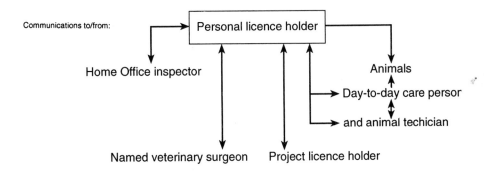

4. **The named day-to-day care person** is usually the senior animal technician within a unit. This person is responsible for ensuring that the animals are kept in the conditions stated in the Home Office Code of Practice and that the animals' environmental conditions are recorded. The day-to-day care person should be familiar with the Certificate of Designation and the project licences in use and must ensure that all the animals are checked daily. The day-to-day care person must notify the personal licensee or arrange for the care or destruction of an animal whose health or welfare gives rise to concern.

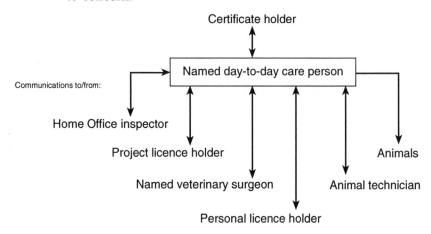

5. **The named veterinary surgeon** should monitor the health status of the animals regularly and advise licensees on matters relating to animal welfare, such as methods of pain recognition and alleviation, surgical technique, and so on. He or she is responsible for directing the use of controlled and prescription-only medicines in the animals and should be familiar with the project licences in use. Like the named day-to-day care person, the named veterinary surgeon who has observed an animal whose health or welfare gives rise to concern must contact the personal licensee or arrange for its care or destruction and is legally responsible to the Home Office for the animals' welfare over and above the responsibilities of the licensees.

   The named veterinary surgeon's responsibilities are laid out in a Code of Practice published by the Royal College of Veterinary Surgeons.

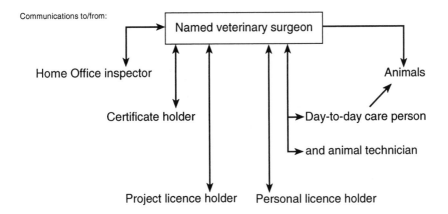

Communications to/from:

6. **The Home Office Inspector** is responsible for ensuring that the work carried out complies with the Animals (Scientific Procedures) Act 1986 and that the terms and conditions of licences are met. Inspectors advise licensees, others who have responsibilities under the Act, and the Secretary of the State for the Home Office.

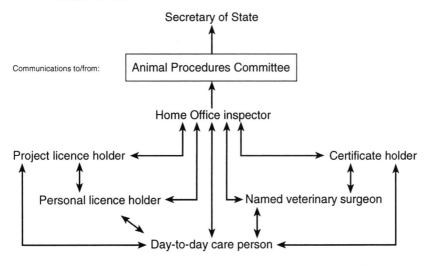

The Animal Procedures Committee advises and reports to the Secretary of State on matters relating to the Animals (Scientific Procedures) Act 1986.

4 Health, safety, and security

EOIN S. HODGSON
University of Oxford Occupational Health Physician and Lecturer in Occupational Health

The hazards and risks of handling animals have almost certainly been appreciated by man since he first became a hunter. This is indicated by prehistoric cave paintings showing death or serious injury during hunting. The subsequent development of animal husbandry and early techniques of research to identify the most useful species to domesticate certainly exposed early man to the hazards associated with animals.

It could be claimed with justification that animal husbandry and its attendant research is a candidate for the title of 'oldest profession' despite claims to the contrary from more controversial trades. Together with the parasitic zoonotic infestations and allergen-acquired chest conditions of the early farmer, animal handlers can claim to be amongst the earliest workers suffering disease associated with their occupations. Life then being brutish and short, however, it is unlikely that they suffered significantly from the chronic effects.

THE HAZARDS

Exposure to animals, and the equipment, chemicals, and infectious agents found in laboratories may under certain circumstances cause injury or illness and therefore adverse contact with them can be said to constitute a hazard. The risk of working with such hazards, however, depends upon how the hazard is contained or secured, the competence of the people involved, and the precautions taken to avoid illness or injury and to protect the workers.

The major potential hazards to animal handlers can be divided into three groups:

1. Allergy
2. Infection
3. Injury

It would be a fortunate animal handler indeed who did not suffer at some time in their career from at least one of these occupationally acquired afflictions. The problems encompass the spectra of both occupational health and occupational safety, with much overlap in methods of recognition, monitoring, control, and surveillance. It is essential for the well-being of the employee in question that both the occupational health adviser (physician or nurse) and safety adviser liaise closely at all stage of investigation and subsequently inform management on the necessary action required to control risk factors.

THE LAW

In most national jurisdictions general action is required in matters connected with health and safety at work. For example, section 2 of the Health and Safety at Work Act 1974 (United Kingdom) requires employers to ensure 'so far as is reasonably practicable . . . the health, safety and welfare at work of all their employees'. Significantly, this legislation extends the duty to protect the health and safety of non-employees '. . . affected by work activity'. Thus by definition, students, visitors, and contractors are owed a duty of care within the law. European Community directives which are binding upon members of the Community extend this general requirement of care to specific activities including the handling of biological agents, hazardous substances, and genetic modification. These affect animal handlers and researchers. Occupational health and safety personnel will need a detailed knowledge of legislation in order to translate it into action within the workplace. In many jurisdictions animal-related conditions, such as occupational asthma and certain animal mediated infections, must be reported to enforcement agencies and may attract government-funded compensation if 'prescribed' in strict legal terms. There are no internationally recognized classifications of such 'prescribed' occupationally related diseases, but animal-acquired occupational asthma is often eligible for compensation. The affected individual also has recourse to the courts in common law, and considerable sums in compensation may be awarded, particularly where punitive damages are likely to be high as in North America.

Increasingly, insurance companies are requiring very detailed assessment of potential hazard and risk, with appropriate control procedures in place, before they will accept the risk. Premiums are likely to rise as claims for damages increase.

Employees are becoming increasingly aware that control and monitoring methods are available and there is some evidence that technicians, particularly these involved with animal handling, are making claims for damages especially when their careers may be blighted by occupationally acquired disease. So far, research workers are reluctant to make such claims, and indeed there are anecdotal reports from occupational physicians of a surprising lack of awareness of the potential effects of their work on their health. Given the fact that some affected employees will be precluded from continuing with their career for legal reasons, it is likely that this cosy situation will not last much longer.

Management is required to respond specifically to employee risk factors. Decisions about suitability for work, both at pre-employment and during employment, require difficult conclusions to be made about future exposure. Academic organizations are not (under most jurisdictions) exempt from health and safety legislative requirements, although many would like to think so. Enforcement authorities and the courts will increasingly demand action where exposures are not controlled adequately, or where individuals continue to be adversely affected.

This chapter discusses the health and safety problems associated with animal handling and suggests some solutions in general terms.

ALLERGY

Confusion over the appropriate acronym for allergy to laboratory animals is common. For the purposes of this section the abbreviation ALA (allergy to laboratory animals) will be used. In some texts the abbreviation LAA (laboratory animal allergy) is preferred.

In 1730, Ramazzini noted, 'Small fragments of dead silkworms as well as locust larvae and caterpillars possess some sort of noxious and corrosive acrimony injurious to the lungs'. This observation by the father of occupational medicine is almost certainly the earliest scientific statement made about what was probably already recognized in those affected trades.

A definition of ALA from Bland *et al.* (1987) states that 'Laboratory animal allergy can be defined as a type 1 immediate allergic reaction mediated through IgE antibody occurring upon exposure either through direct skin contact with or through inhalation of allergen . . . Inhalation of these allergens results in symptoms such as rhinitis, conjunctivitis, and asthma-like symptoms'. The situation however is not quite so simple. Certainly the bulk of ALA may be of the immediate type, and rhinitis (or hayfever-like symptoms) is extremely common in those affected, but other sensitizing mechanisms are also implicated in ALA, including IgG-mediated allergy, which may only occur 1 to 12 hours after exposure, and non-IgE-mediated symptoms involving other immuno globulins and giving rise to specific pathology, such as extrinsic allergic alveolitis (EAA), rather than rhinitis, conjunctivitis or asthma.

It has been estimated that between 9 and 30 per cent of animal

handlers will exhibit one or more symptoms of allergy, the commonest being rhinitis or hayfever-like symptoms (sneezing and running
nose), but a variety of other symptoms are commonly recorded,
including:

1. Rhinitis (sneezing, running/blocked nose).
2. Conjunctivitis (stinging and running eyes often with associated injection of the conjunctiva and local swelling of the
   eyelids).

These two symptom groups are the commonest associated with
animal allergy.

3. Skin effects.
   (a) Urticaria ('hives' or 'nettlerash').
   (b) Wheals, that is raised red areas around bites and scratches.
   (c) Eczema, particularly on the backs of the hands and
       occasionally the face.
4. Asthma.
5. Extrinsic allergic alveolitis.
6. Anaphylactic reactions, which may be life-threatening.

Many workers have noted that the most common form of animal
allergy is rhinitis, and in many of those affected there is no pro-
gression to other symptom groups. Atopic individuals (that is those
producing IgE antibody following exposure to allergens) have been
shown by some workers (Kibby *et al.* 1989) to be more likely to
develop ALA, but others have questioned this hypothesis.

Where ALA occurs it is likely to manifest itself within 2 years of
first exposure to the species. Late onset of symptoms after many years
exposure, however, is also recorded.

The most common symptom group is rhinitis and those who
are affected may not make the connection with their exposure,
particularly if they suffer from concomitant hayfever. The symptoms
may be mild, require no treatment, and are unlikely to be reported
to medical advisers. Even with the onset of conjunctivitis and skin
symptoms the connection may not be made by the individual, and
unless symptoms of cough and wheezing with shortness of breath
occur, no investigation will be carried out.

Where asthmatic symptoms do develop they may be considered
as nothing more than an increase in the asthma already established
in the individual, who will probably be atopic. Many asthmatics
whose condition is well controlled by regular therapy (including
inhaled steroid) may notice nothing more than a slight increase in
the demand for bronchodilation, and unless the condition becomes
more of a nuisance, they are unlikely to seek medical intervention
or change of treatment.

Extrinsic allergic alveolitis is quite different from asthma, but nevertheless an allergic phenomenon. Often unrecognized until breathlessness causes considerable disability, it is certainly under-diagnosed. The acute form may resemble an influenza-like illness with sweating, cough, generalized muscle pains, and malaise. Recovery usually occurs quite rapidly, but re-exposure may cause recurrence of the symptoms. The chronic condition may be accompanied by intermittent cough and influenza-like illnesses with increasing breathlessness. The condition is probably a good deal less common than asthma, but misdiagnosis is the rule rather than the exception.

Anaphylactic shock with bronchospasm, angio-oedema, and hypotension may be life-threatening and may occur in animal handlers who are bitten by snakes or stung by insects, particularly bees and wasps. Although rare, such reactions may cause death because effective treatment cannot be instituted soon enough. Death may be due to laryngeal oedema and subsequent hypoxia. Those likely to be affected must avoid exposure and this can, of course, have devastating effects on professional researchers.

Given the fact that allergy to laboratory animals is common, health surveillance is required for all those handling animals. Surveillance should begin with pre-employment assessment, and those identified as having major problems should be guided into other careers. Surveillance will include symptom questionnaires and self-reporting of symptoms to the occupational health advisers. Laboratory estimation of immune globulin levels and perhaps radio-allergo sorbent testing (RAST) may be required. Where symptoms are progressing despite appropriate control measures, individual advice on continuation of exposure will be required. Occasionally those affected should be advised to avoid exposure and seek an alternative career.

Causes of allergy

The most common cause of allergic symptoms in animal handlers is probably urinary protein with hair, fur, dander, saliva, and serum also implicated. Contact with the droppings from birds, locusts, and crickets may also cause allergy. Methods of control should seek to reduce aerosol production which will require appropriate technology including local exhaust ventilation and specialized cleaning techniques for cages.

Anecdotally, the rat is considered to be the most likely to cause allergy, but some workers suggest the cat, guinea-pig, and rabbit are stronger candidates (Bland *et al.* 1987).

Control

Although most allergens will produce symptoms at very low levels of exposure, there is some evidence that the control of levels may reduce the amount of animal allergy and its severity. The hierarchy of control would classically require elimination of the allergen, and, where that is not possible, a series of measures designed to reduce exposure. The United Kingdom Control of Substances Hazardous to Health Regulations Approved Code of Practice (Health and Safety Commission 1993) suggests the following:

(a) Totally enclosed processes and handling systems.

(b) Plant or processes or systems of work which minimize the generation of, suppress, or contain, hazardous dust, fumes, microorganisms, etc., and which limit the area of contamination in the event of spills and leaks.

(c) Partial enclosure with local exhaust ventilation.

(d) Local exhaust ventilation.

(e) Sufficient general ventilation.

(f) Reduction of numbers of employees exposed and exclusion of non-essential access.

(g) Reduction in the period of exposure for employees.

(h) Regular cleaning of contamination from, or disinfection of, walls, surfaces, etc.

(i) Provision of means for safe storage and disposal of substances hazardous to health.

(j) Prohibition of eating, drinking, smoking, etc., in contaminated areas.

(k) Provision of adequate facilities for washing, changing, and storage of clothing including arrangements for laundering contaminated clothing, to which may be added the requirement for personal protective equipment including respiratory protective equipment which is only acceptable where other measures are already in place.

The Education Services Advisory Committee in the United Kingdom has produced two documents which suggest specific ways of controlling animal allergens (Health and Safety Commission 1990, 1992), in particular it concentrates on adequate general ventilation, specific ventilation for animal rooms, and specially designed cleaning systems to reduce the likelihood of aerosol production. Personal protective equipment (PPE) is essential for all people working with laboratory animals or entering the facilities, and correct disposal of or laundering of this is also required.

If engineering controls are unable to reduce allergen exposure to acceptably low levels, respiratory protective equipment (RPE) will be required. Such equipment should be carefully chosen, and training given to individuals on fitting. Compliance with national standards is mandatory. Those who are particularly allergic, (with symptoms of asthma, for instance), may require air-fed visors or helmets with high-efficiency filtration. However, for most individuals, disposable dust respirators, which must be discarded after usually 4–8 hours use, should give sufficient protection to control symptoms.

The combination of appropriate control measures and adequate, specific and sensitive health surveillance should enable most workers affected by animal allergy to continue working. There will always be those who should be advised to avoid further exposure. Careful consultation will be required on a case by case basis with the occupational health adviser, and redeployment of those severely affected should be arranged. Indeed, in some jurisdictions such re-deployment is mandatory.

INFECTION

The hazards from microorganisms and parasites to animal handlers are well known. The world-wide farming community is only too well aware of the wide variety of animal-induced diseases which can be transferred to humans, ranging from common gastrointestinal infections (from *Campylobacter* and *Salmonella* spp.) to rare, but life-threatening haemorrhagic fevers (Marburg and Hantaan viruses), Weil's disease (*Leptospira* spp.) and Simian B disease (from *Herpesvirus simiae*). Over 150 zoonoses have been recognized of varying significance to humans, and hookworm infestation is probably the most common occupationally-acquired disease in the world, being universal amongst the barefoot farmers of developing countries.

There is remarkably little good scientific data on the incidence of zoonoses amongst animal handlers, but anecdotal evidence from the UK Education Services Advisory Committee (Health and Safety Commission 1992) is listed in Table 4.1. Ringworm is a common zoonosis in farming communities and in those animal facilities involved with larger domestic animals including horses, cattle, and sheep. Rabies is a potential hazard for animal handlers, but quarantine methods and vaccination of humans (and perhaps animals) renders this an unlikely zoonosis for most animal handlers in the United Kingdom.

Researchers working in the field may be exposed to more life-threatening infections than those working in laboratories. These include the haemorrhagic fevers (such as Marburg disease carried by some monkeys, and Lassa fever from rodent urine), which are potential hazards for those working in the affected parts of Africa. Control methods include careful veterinary investigation of such animals, and strict adherence to safety precautions, such

Table 4.1
Examples of zoonoses in animal facilities

Organism	Animal source	Human disease	ACDP hazard group
Campylobacter	Various	Campylobacteriosis	2
Chlamydia psittaci	Sheep	Ovine chlamydiosis	2
	Birds	Avian chlamydiosis	3
Coxiella burnetti	Sheep and cattle	Q fever	3
Cryptosporidium	Sheep and cattle	Cryptosporidiosis	2
Hantaan virus	Rat	Korean haemorrhagic fever	3
Herpesvirus simiae	Simians	Simian B disease	3
LCM virus	Mouse	Lymphocytic choriomeningitis	3
Leptospira	Rat	Weil's disease (leptospirosis)	3
Microsporum and *Trichophyton*	Various	Ringworm	2
Salmonella spp.	Various	Gastroenteritis	2
(*S.typhi* and *paratyphi*)	(Fruit-eating bat)	(Typhoid and paratyphoid)	(3)
Shigella spp.	Primates	Bacillary dysentery	2
(*Shigella dysenteriae*, type 1)			(3)
Streptobacillus moniliformis	Mainly rat and mouse	Rat bite fever (Haverhill fever)	2
Toxoplasma gondii	Cat	Toxoplasmosis	2
Rochalimaea henselae*	Cat	Cat scratch disease (benign lymphoreticulosis)	3

After ACDP, advisory committee on dangerous pathogens 1990; (3) = 2 or 3. * Zangwill et al. 1993.

as the wearing of protective clothing, and the use of chemical restraint.

Particular hazards from simians

The Medical Research Council (UK) (Medical Research Council 1990) has made particular suggestions about infectious hazards from simians to animal handlers with particular reference to simian herpesvirus (Simian B-virus, Herpesvirus simiae), and simian retroviruses (Whitley 1990).

Simian herpesvirus

This occurs in Old World monkeys (but not the great apes), and has caused at least 22 cases of disease in humans, most of which were fatal (Medical Research Council 1990). Control relies on using B-virus-free animals and tissues, and not introducing individuals of unknown status to established colonies. Avoidance of bites and spitting by careful handling and welfare techniques is also required (Whitley 1990).

Simian retroviruses

Simian immunodeficiency viruses (simian immunodeficiency virus—SIV, and simian T-lymphotrophic lentivirus—STLV) are related to human immunodeficiency virus, but it is not yet known whether humans can be affected by SIV or STLV. Retroviruses are excreted by simians via saliva and urine, but so far no cases of transmission between monkeys and man have been reported. Appropriate containment levels will be required for work with immunodeficiency viruses whether simian or not, and the Advisory Committee on Dangerous Pathogens (1990) requires a high level of containment for such work.

Control methods

Animals should always be purchased from accredited breeders, and reports on the microbiological status of the animals should be requested by purchasers. Regular screening of animal stocks should occur once animals are purchased. Veterinary advice will identify the most likely pathogens for screening (see Chapter 7).

Hazardous materials

All body fluids from animals should be considered potentially hazardous, and appropriate precautions taken to avoid contamination of the skin, mucous membranes, or blood of the animal handler.

Introduced pathogens and tissues

Deliberately introducing pathogens into animals for study causes special problems for animal handlers. The rigid application of containment facilities is mandatory in these circumstances and, where appropriate, vaccination of the animal handler will be required (Advisory Committee on Dangerous Pathogens, 1993). If human material is being used it should be screened, where possible, particularly for hepatitis B, hepatitis C, and human immuno-deficiency viruses. Special precautions including flexible film isolators or filtertop cages may be required within specialized containment areas.

Health surveillance

It is difficult to identify those likely to be affected by micro organism contamination. In some zoonoses an acute illness will occur, and will be identified by investigation. However, it is reasonable to obtain pre-employment serum samples for storage from researchers working with potentially life-threatening human viruses in animals, for example, hepatitis B and HIV. Some jurisdictions require regular sample taking for storage (e.g. at 6 monthly intervals), but the logic of this is difficult to justify. However, following percutaneous contamination injuries with infected tissue or blood it is usual to take the assaulted individual's blood for storage. Hepatitis B vaccination is mandatory for those working with the virus under any circumstances.

Pre-employment health assessment of those exposed to pathogens in animals may require the exclusion of immune-suppressed workers from exposure to pathogens. The possibility of opportunistic infection in such individuals may be considered too high a risk.

Those animal handlers working with vaccinia and related pox-viruses will need special advice, and the Advisory Committee on Dangerous Pathogens has suggested that only those working with monkeypox virus must be vaccinated, but other exposures might require vaccination following case by case analysis. Contraindications for vaccination include pregnancy, a history of eczema or other active skin disease, and immune deficient states (Advisory Committees on Dangerous Pathogens and Genetic Modification 1990). The employer must make the decision as to whether those who should be vaccinated, but cannot be, may proceed to work with poxviruses.

The potentially major risks of contamination with microorganisms during animal handling are obvious. However, control measures including selection of appropriate animal containment facilities, good handling techniques and, where available, human vaccination, will considerably reduce the likelihood of actual infection. A high index of suspicion for potential adverse effects and early investigation of symptoms is required.

INJURIES

Animal species (including insects) like humans, tend to resent assault, and react defensively. Many large domesticated animals may, by their very bulk, damage humans unintentionally. For example, the horn of a cow may pierce the stockman purely by turning of the head. Smaller species, such as mice and hamsters, may bite if poorly handled, and even from apparently clean laboratory animals these bites can become infected, with a variety of commensal organisms including *Streptococci* and *Staphylococci* causing potentially major infections in humans. It is axiomatic that all injuries should be reported, and most health and safety jurisdictions require detailed information on such injuries and appropriate action by the employer.

Animal handlers may sustain a wide variety of injuries from their charges, some of which may be life-threatening, such as:

(a) Soft tissue injuries (bruising) and fractures from stamping, butting and crushing.

(b) Bites
    Non-toxic, especially from pigs, dogs, and small mammals.
    Toxic, in particular snake, possibly lizard and spider bites.

(c) Stings including fish, such as weavers, stone fish, etc., coelenterates, echinoderms (starfish and sea-urchins), mollusca (cone shells), and arthropods (including bees, wasps, hornets and scorpions).

Given the wide variety of potential methods of injury, it follows that damage to the sufferer may be very mild, as in the case of some insect bites and stings, or so severe as to cause death as in major injuries from large animals, or bites from particularly venomous snakes (Hydrophiidae and Atractaspididae).

It is mandatory to have a rapid line of communication between handlers of venomous species and medical help. Where available, antivenoms should be obtained quickly, and experts in the assessment and treatment of envenomings should be identified.

Prevention

The prevention of injuries is difficult due to the unpredictable nature of many species. The male of all species can be much more aggressive than the female, but all species with young will react defensively to protect their offspring. Snake and insect bites usually occur when animals are surprised or disturbed by loud noises. Careless handling techniques will cause stress to the animal and without appropriate protective equipment (including where necessary heavy duty leather or metal protected gloves) can lead to injury (see Chapter 9).

It is crucial that the animal handler understands the species he/she is working with, is aware of their habits and likely defensive tactics, and has been trained to approach the species correctly. Specifically designed containment and caging facilities are required, particularly with large animals when specimens are to be obtained. In the wild, anaesthetic dart techniques may be necessary to approach the animals safely.

First aid facilities should be available in all animal handling premises and individuals trained in first aid techniques to the appropriate national standard are required. They should have information on the nature of the animals and the likely injuries they could cause and, in the case of venomous species, particularly snakes, they will need training in specific first aid methods including standard procedures for containing the spread of venom. This may, for instance, include arterial tourniquet methods which are inappropriate in other forms of first aid. Clear lines of communication to expertise via the local emergency facility are also required.

LONE WORKING

Where there are identified hazards from occupations there are particular risks in working alone. This is especially so of animal handlers, particularly in the field, with the possibility of severe injury or potentially life threatening envenomization or anaphylaxis. In all cases workers must have a rapid means of communication to someone who can respond on their behalf. This is naturally much easier in a purpose-built facility in the middle of a centre of population, but may be impracticable in remote locations. Such workers accept the (theoretically) major risks associated with such work, but in order to make this judgment they must be aware of the potential risks and make their own assessment in conjunction with their employer. Some lone workers carry alarms which are activated if they change position rapidly, and in certain areas carrying personal radios or portable telephones is a distinct possibility which should be considered. They should ensure that somebody knows where they are and not deviate from their plans however tempting may be an alternative scenario. Checking into and out of animal facilities with security-coded entrances is now common and an added reassurance for the lone worker.

MISCELLANEOUS HEALTH AND SAFETY HAZARDS AND RISKS

Animal handlers are exposed to a wide variety of other occupational hazards and risks similar to those which occur elsewhere, for instance:

Chemical hazards

1. Disinfection agents may cause irritation of skin and mucous membranes, including the respiratory tract and/or sensitization of the skin and lungs. Appropriate handling techniques are described by the manufacturers and should be available.

2. Anaesthetic agents, particularly those used for large animals (e.g. etorphine, Immobilon) may require specific antidotes (e.g. diprenorphine, Revivon or naloxone, Narcan) to be drawn up with a second person available in order to inject them rapidly. Other drugs may cause sensitization if carelessly handled (e.g. prostaglandins) and some are possibly carcinogenic (e.g. xylene, urethane) and require special handling techniques which require training and containment methods.

3. General laboratory chemicals may include mineral acids, alkalis, phenol derivatives, aromatic solvents, and pesticides such as organophosphate insecticides, which have their own hazards well described in other texts (Royal Society of Chemistry 1992).

Electrical hazards

Animal facilities require electrical apparatus which should be inspected on a regular basis. Care should be taken when using sharp instruments close to a power cable, and plugs and sockets should not be contaminated with animal products, or allowed to get wet during cleaning. Most jurisdictions have specific requirements to control the use of electricity at work and these should be adhered to.

Radiation

1. Ionizing radiation. Appropriate codes of practice should be followed. The worker may be required to undergo regular health surveillance by an approved doctor.

2. Non-ionizing radiation. Ultraviolet light, microwaves, and lasers all possess specific hazards which require action by employers and employees. Careful control is required for all such emissions. Class 4 lasers may pose a specific fire hazard as well as potentially causing damage to the eyes and skin even from reflection.

Sharp instruments

Sharps injuries are common in animal handlers, partly due to the unpredictability of their charges. Percutaneous injury poses the risk of blood-borne transmissible disease or local infection and careful use of sharp instruments is required. Disposal in appropriate containers is mandatory. Health and safety legislation may require containers constructed to a specific standard with subsequent disposal by special means including incineration in dedicated units. In many cases, for instance, it is contraindicated to resheath needles once used, but in other circumstances resheathing is the safest way of controlling the potential hazard.

Local laboratory practice in writing ('local rules') should identify methods of handling and disposal of such instrumentation.

Waste disposal

The cleaning of cages and the disposal of animal waste is a specialized technique which needs to avoid the formation of aerosols of animal products in the environment. Such aerosols may settle only very slowly despite appropriate ventilation and will pose a hazard, particularly in terms of allergen content. Animal waste should be collected carefully, bagged in appropriate containers, labelled and disposed of according to national waste disposal regulations. It must not be treated like household waste. Carcasses and tissue specimens from animals may require double bagging in heavy gauge plastic bags of particular colour or labelling and should be collected for incineration by trained personnel.

Certain material associated with animal experimentation will require autoclaving prior to disposal by an appropriate route.

EMPLOYEE SECURITY

The very real problem associated with animal rights groups world wide poses a considerable threat to the personal security of all those who work with laboratory animals. Animal facilities should be especially secure and will require high levels of entrance/exit control. The building should be designed so that breaking and entering is difficult or impossible and high levels of alarm systems connected to local police facilities are indicated.

Individual employees may wish their identity to remain confidential. In particular, the home addresses and telephone numbers of animal handlers should be kept completely confidential in secure storage. Such employees should not carry identification which indicates that they are animal handlers unless absolutely necessary.

However, it is advised that animal handlers (along with many other employees) should carry a card indicating the nature of their work in general terms. This should increase the 'index of suspicion' of a doctor or health facility to whom they are unknown if taken ill perhaps in a remote or strange location. This HAZCARD should be worded carefully and should not identify the individual's address, telephone number or next of kin. A suggested form of card is shown in Figure 4.1, which attempts to get round the problem of identifying animal handlers.

Side 1

ACME Research Institute   **HAZCARD**
Occupational Health Service

Name...
may be exposed to environments/substances hazardous to health, some of which are defined in the COSHH Regulations 1988. Health records of exposure are kept at the Occupational Health Service. Advice on possible adverse effects may be obtained from:
Dr A.N. Other (tel no)
or the person named by the card holder

Side 2

GP..
Tel..

Substances hazardous to health include: chemicals, commercial preparations, microorganisms (and zoonoses), dusts, fibres, vapours, gases, allergens, botanical toxins. Environments also pose health hazards. The card holder will discuss the exposure type with you. *Please have a high index of clinical suspicion if symptoms/signs occur in this worker who is your patient.
Dr A.N. Other   Occupational Health Physician

4.1 HAZCARD.

5 Humane methods of killing

The Animals (Scientific Procedures) Act 1986 lists in Schedule 1 standard methods of humane killing (see Figure 5.1). Neither a project nor personal licence is required to carry out the methods on this list for the animals as indicated. However, it is required that the person carrying out the killing is competent to do so without causing distress to the animals involved.

If a method of euthanasia is to be used which is not in Schedule 1, or if one of the listed methods is to be used for an animal for which it is not deemed appropriate, then it must be authorized on both the project and personal licences before it can be carried out.

Killing an animal is always an unpleasant task. It is important that the animal is handled carefully and competently without causing it distress (see Chapter 9), until unconsciousness has occurred. This should happen rapidly and be swiftly followed by cardiac and respiratory arrest.

If the animal is frightened it may exhibit total immobility or it may show behavioural responses such as vocalization, struggling, urination, defaecation, anal sac emptying, and muscle tremor. This fear can be communicated by sound or smell causing distress to the other animals. Some of these responses may be exhibited by the animal when it is unconscious but before death occurs, so it is most important that animals are removed to another room away from the group before they are killed.

When choosing a method of euthanasia, consider the following points:

1. Death must occur without producing pain.
2. The time required to produce loss of consciousness must be as short as possible.
3. The time required to produce death must be as short as possible.
4. The method must be reliable and non-reversible.

Method	Animals for which appropriate
A. *Animals other than fetal, larval, and embryonic forms*	
1. Overdose of anaesthetic suitable for the species:	
(i) by injection	(i) All animals
(ii) by inhalation	(ii) All animals up to 1 kg bodyweight except reptiles, diving birds, and diving mammals
(iii) by immersion	(iii) Fishes Amphibia up to 250 g bodyweight
(Followed by destruction of the brain in cold-blooded vertebrates and by exsanguination or by dislocation of the neck in warm-blooded vertebrates except where rigor mortis has been confirmed)	
2. Dislocation of the neck	Rodents up to 500 g bodyweight other than guinea-pigs
(Followed by destruction of the brain in fish)	Guinea-pigs and lagomorphs up to 1 kg bodyweight Birds up to 3 kg bodyweight Fishes up to 250 g bodyweight
3. Concussion by striking the back of the head	Rodents up to 1 kg bodyweight Birds up to 250 kg bodyweight Fishes
(Followed by exsanguination or dislocation of the neck in rodents and birds and destruction of the brain in fish)	
4. Decapitation followed by destruction of the brain	Cold-blooded vertebrates
5. Exposure to carbon dioxide in a rising concentration using a suitable technique followed by exsanguination or by dislocation of the neck except where rigor mortis has been confirmed	Rodents over 10 days old up to 1.5 kg bodyweight Birds over 1 week old up to 3 kg bodyweight
B. *Fetal, larval, and embryonic forms*	
1. Overdose of anaesthetic suitable for the species:	
(i) by injection	(i) All animals
(ii) by immersion	(ii) Fishes Amphibia
2. Decapitation	Mammals

5.1 Schedule 1: Standard methods of humane killing.

5. There must be minimal psychological stress on the animal.

6. There must be minimal psychological stress to the operators and any observers.

7. It must be safe for personnel carrying out the procedure.

8. It must be compatible with the requirements of the experiment.

9. It must be compatible with any requirement to carry out histology on the tissues.

10. Any drugs used should be readily available and have minimum abuse potential.

11. The method should be economically acceptable.

12. It should be simple to carry out with little room for error.

SCHEDULE 1 METHODS
Overdose of anaesthetic
Injection
With modern injectable agents used in combination to achieve balanced anaesthesia (see Chapter 12) there is generally a fairly wide safety margin, and indeed with some anaesthetics it is actually quite difficult to kill the animal using a reasonable injection volume. When administering an anaesthetic, one is seeking to avoid death rather than to cause it. When carrying out euthanasia it is therefore important to select the agent carefully. The drug commonly used to carry out euthanasia is pentobarbitone, which indicates its unsuitability as an agent for safe, reversible anaesthesia. It is usually presented as a 20 per cent solution (200 mg/ml) and is administered at a dose of at least 140 mg/kg. It is preferable to give it intravenously for the most rapid action but in the smaller species it is generally administered intraperitoneally, using a suitably sized needle (see Chapter 10).

In the larger animals, if it is not possible to locate a vein in the conscious animal, it should be sedated with an agent administered by an easier route. When the sedative has taken effect and the animal can be handled more easily, the pentobarbitone can be administered intravenously to kill it quickly and humanely. Pentobarbitone must not be given intramuscularly as it is very irritant to the tissue and this will cause pain. Intrathoracic injections in the conscious animal are also painful, and an intracardiac injection of pentobarbitone should only be attempted after the animal has been rendered unconscious by some other agent (see Chapter 12 on anaesthesia).

Inhalation
To carry out euthanasia by overdose of inhalation anaesthetic, the animal is placed in a suitable induction chamber with the vapour.

It is important that the animal is physically separated from the liquid agent since volatile anaesthetics are very irritant to mucous membranes. Generally, the agent is either poured on to a pad of cotton wool which is placed in the chamber but separated from the animal by a grid, or the agent is piped into the chamber from an anaesthetic machine fitted with a vaporizer. Depending on the agent chosen and the concentration used, unconsciousness occurs quite rapidly but death may take rather longer. It is therefore important to ensure that the animal is left in the chamber for long enough, and to confirm that it is dead, since if it is removed too soon and allowed to breathe room air, it may recover. The chamber should be designed so that the animal can be easily observed, and so that it can be easily cleaned between batches of animals to remove all traces of urine and faeces, which contain pheromones by which the animals communicate stress. Placing the animals on disposable paper which can be replaced each time, is a simple way of controlling this potential stress factor. The use of agents such as ether or chloroform for this purpose is not considered humane, because of the intense irritation caused to the respiratory system before unconsciousness occurs which leads to a sensation of choking. Chambers of this type should always be used in an extraction cabinet or with a suitable ducting system to keep the volatile agent away from the operator.

Immersion

For fishes and small amphibia euthanasia can be achieved by percutaneous administration of an agent such as tricaine methane-sulphonate (MS 222). As for inhalation euthanasia in the rodent, care must be taken to ensure that the animal is left in the solution for an adequate length of time for death to occur before it is removed.

Physical methods of euthanasia

Physical methods of euthanasia have the disadvantage that they are usually distasteful to the person carrying them out, and this leads to a tendency to be rather hesistant. However, human feelings should not be allowed to influence the choice of the most humane techniques and if carried out properly these methods are often less distressing to the animal, since death is very quickly achieved. Manual dexterity and an ability to handle the animal confidently are essential to minimize any apprehension. They should be practised first on dead animals, after careful observation of the method carried out by a competent and experienced person.

Dislocation of the neck

For the smaller rodents, put the animal on a surface on which it can grip, place a pencil or similar object firmly across the back of the neck, take a firm grasp around the hindquarters or the tail and pull sharply. The neck will be dislocated and the animal will die instantly. For larger rodents and lagomorphs, hold the body firmly in one hand, the head in the other and pull sharply with a rotating action.

Concussion

As with dislocation of the neck, confidence in handling the animal and manual dexterity are required to carry out this method of euthanasia without causing distress to the animal. It should be supported by the hindquarters and the body swung downwards such that the back of the head comes sharply into contact with a hard surface such as a work bench. Following concussion, which will only render the animal unconscious, death must be ensured by dislocation of the neck or by opening the body cavity and severing the major blood vessels so as to exsanguinate the animal.

Decapitation

This is only suitable for cold-blooded vertebrates or mammals in fetal form. The animal is held firmly and the head removed using a pair of stout, sharp, scissors. For cold-blooded vertebrates the brain must immediately be destroyed by pithing (inserting a rod into the cranium and moving it about to destroy the brain tisssue).

Exposure to carbon dioxide

It is important to realize that the concentration of carbon dioxide used for euthanasia must be rising. The animal is placed in a chamber which, like the one used for overdose of inhalation agent, should be easy to clean and give a clear view of the animals inside. The flow of carbon dioxide is turned on and gradually rises, displacing the air. The animals gradually become hypoxic and lose consciousness. As the hypoxia increases, so they die. After use the chamber must be inverted to tip out all the residual carbon dioxide since it is heavier than air and will sink to the bottom of the chamber, although of course it cannot be seen. If this is not done and animals are put into the chamber when the concentration is already high, or if the concentration is increased too rapidly, they will exhibit respiratory distress as they fight for breath.

Disposal of carcasses

After carrying out any method of euthanasia it is vital to check that the animal is really dead before disposing of the body. It can either be left until rigor mortis has set in, or death can be ensured by the severing of major vessels or by dislocation of the neck. All carcases from research institutes are classed as clinical waste and as such must be disposed of correctly. There will be local rules in the work area relating to the method of disposal and these must be followed.

6 Introduction to laboratory animal husbandry

GENERAL CONSIDERATIONS

The laboratory is far from being the natural environment for the animals that are kept there. Laboratory animals do not have the freedom to move away from adverse conditions, to search for food and water, or to find a better nesting site. All their physiological and behavioural needs must be supplied by the laboratory environment. These needs are not optional: they must be catered for or poor health will result.

Caging or housing needs to keep terrestrial mammals dry, clean, and warm. There must be adequate room to allow a normal range of movement, without overcrowding, and there must be free access to food and water. Guidelines on minimum cage sizes and stocking densities are published by the Home Office (see Table 6.1). Cages must be made from harmless material which is easily cleaned and sterilized, and must be able to withstand attempts to escape. Some animals are particularly adept at chewing through plastic cages or removing food hoppers to escape.

Bedding materials may be used to provide warmth and comfort, and to improve sanitation. The bedding must be harmless, i.e. non-toxic and non-irritant, hygienic, absorbent, easily disposable, cheap and readily available, and easy to use. Bedding should be changed as often as is required to maintain a clean environment, and the frequency will vary depending on the species.

Several factors contribute to the quality of the environment. The temperature, humidity, ventilation, and lighting intensity must be maintained at a level appropriate for the species. The lighting level should be monitored both in the room and in the cages. It is likely to be dimmer in the cages than in the room, but light in cages on the top row of a rack may be significantly brighter than in other cages, which can be a problem for albino animals with little pigment in the retina. It is advisable to cover the top row so the whole population is in as uniform an environment as possible, or such variable factors may

Table 6.1
Cage and pen dimensions for the housing of laboratory animals
(taken from the Home Office Code of Practice)

| Weight of animal | Minimum floor area (sq cm per animal) | | Minimum height (cm) |
	When housed in groups	When housed singly	
Small mammals			
Mouse			
up to 30 g	60	200	12
over 30 g	100	200	12
Rat and gerbil			
up to 50 g	100	500	18
50 to 150 g	150	500	18
150 to 250 g	200	500	18
250 to 350 g	250	700	20
350 to 450 g	300	700	20
450 to 550 g	350	700	20
over 550 g	400	800	20
Hamsters			
up to 60 g	80	300	15
60 to 90 g	100	300	15
90 to 120 g	120	300	15
over 120 g	165	300	15
Guinea-pig			
up to 150 g	200	700	20
150 to 250 g	300	700	20
250 to 350 g	400	900	20
350 to 450 g	500	900	23
450 to 550 g	600	900	23
550 to 650 g	700	1 000	23
over 650 g	750	1 250	23
Rabbit			
up to 2 kg	1 300	2 000	40
2 to 4 kg	2 600	4 000	45
4 to 6 kg	3 300	5 400	45
over 6 kg	4 000	6 000	45

Table 6.1 (cont.)

Weight of animal	Minimum floor area (sq cm per animal)		Minimum height (cm)
	When housed in groups	When housed singly	
Ferret and mink			
up to 800 g	1 500	2 250	50
over 800 g	3 000	4 500	50
Cat			
up to 3 kg	3 300	5 000	50
over 3 kg	5 000	7 500	80
Dog			
up to 5 kg	10 000	45 000	150
5 to 10 kg	19 000	45 000	150
10 to 25 kg	22 500	45 000	200
25 to 35 kg	32 500	65 000	200
over 35 kg	40 000	80 000	200

Farm animals and *equidae*

Species and weights	Minimum floor area (sq m per animal)		Minimum length of feed rack or trough per head (m)
	When housed in groups	When housed singly	
Pig			
up to 30 kg	1.0	2.0	0.20
30 to 50 kg	1.3	2.0	0.25
50 to 100 kg	2.0	3.0	0.30
100 to 150 kg	2.7	4.0	0.35
over 150 kg	3.75	5.0	0.40
adult boar		7.5	0.50
Sheep and goat			
up to 35 kg	1.3	2.0	0.35
over 35 kg	1.9	2.8	0.35

Table 6.1 (*cont.*)

Weight of animal	Minimum floor area (sq cm per animal)		Minimum length of feed rack or trough per head (m)
	When housed in groups	When housed singly	
Cattle			
up to 60 kg	1.5	2.2	0.30
60 to 100 kg	1.6	2.4	0.30
100 to 150 kg	1.9	2.8	0.35
150 to 200 kg	2.4	3.6	0.40
200 to 400 kg	3.8	5.7	0.55
over 400 kg	5.3	8.0	0.65
adult bull		16.0	0.65
Horses, donkeys and crossbreds height at withers			
up to 147 cm	12		
148 to 160 cm	17		
over 160 cm	20		

Non-human primates

Weight of animal	Minimum floor area (sq cm per animal)		Minimum height (cm)
	When housed in groups	When housed singly	
up to 700 g	1 350	2 500	80
700 to 1400 g {	2 500	5 000	100
	2 000*		150
1400 to 4000 g	6 000	6 000	100
4000 to 6000 g	8 000	8 000	110
6000 to 9000 g	14 000	14 000	150

* For arboreal monkeys in groups when they are held in taller cages.

Table 6.1 (*cont.*)
Birds

| Species and weights | Minimum floor area (sq cm per animal) | | Minimum height (cm) | Minimum length of feed trough per bird (cm) |
	When housed in groups	When housed singly		
Chicken and duck				
up to 300 g	250	350	30	3
300 to 600 g	470	700	40	7
600 to 1200 g	830	1250	50	10
1200 to 1800 g	950	1450	50	12
1800 to 2400 g	1200	1700	55	12
over 2400 g	1900	2800	75	15
Quail				
up to 150 g	250	350	20	4
150 to 250 g	250	400	25	4
Pigeon				
	800	1225	35	5

affect the experimental results. Noise levels should be monitored. Animals' ears do not necessarily respond to the same range of frequencies as human ears. The room may seem quiet to people, but there may be sound uncomfortably loud in the ultrasound range which can be detected by rodents. Therefore the intensity and frequency of the noise in the room should be assessed.

It is not sufficient just to cater for the physiological needs of the animals. The animal must have freedom to express as much natural behaviour as possible, including play and social contact if appropriate. The inability to behave naturally is deleterious to health and welfare, and results in animals performing repetitive, abnormal behaviour patterns. This is known as **stereotypy**. Environmental enrichment caters for the behavioural needs of the animals, and improves their health and welfare as much as any other environmental factor. The provision of play articles or bedding may be a simple way to achieve **environmental enrichment**, allowing the expression of normal behaviour, and improving the welfare of the animals (see Figures 6.1, 6.2, 6.9, 6.13, and 6.14).

6.1 These experimental pigs live in an environment that is enriched by the presence of a mud wallow. The animals use the wallow to assist in thermoregulation and there is the opportunity for natural rooting behaviour in the soil.

Some thought should be given to what environment the animal would choose to satisfy all its needs, and as close a match as possible should be provided by the laboratory.

SMALL SPECIES
Mouse
The mouse, *Mus musculus*, is the most commonly used laboratory animal. Many well-defined inbred and outbred strains are available, for which the karyotypes are known. In fact, more is known about the genome of the mouse than any other species, which is one reason for its popularity as a research animal.

Behaviour
Mice are social animals which can live in harmony, in groups with one male and several females, once their hierarchy is established. The effects of pheromones must be taken into account when managing a mouse colony. Pheromones are used to maintain stability in the colony and, if removed by cleaning, fighting will ensue and subordinate mice will be barbered. Leaving a little of the soiled bedding in the new cage will reduce this, as will sprinkling a little baby powder over the mice to disguise their scent. Pheromones act as mediators in communication between mice, and their influence must not be underestimated.

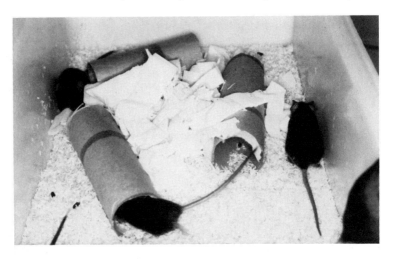

6.2 Even mice benefit from the presence of simple objects used to provide environmental enrichment.

Some of the effects of pheromones are as follows:

— Stress in one mouse causes dispersal of other mice
— Female mice attract male mice, and vice versa
— Lactating females emit pheromones to attract the young
— Foreign females will stimulate aggression by other female mice
— Foreign males will provoke aggression from other male mice, and may cause recently mated females to abort (Bruce effect)
— Coexisting males emit pheromones to reduce aggression within the group, but which cause foreign males to avoid the territory

There are also strain differences in behaviour. For example, male BALB/c mice are particularly aggressive, and fight wounds are common.

Housing
Mice can be housed in conventional units, or in barrier units (See Chapter 7). The latter may be for axenic mice (full barrier environment resulting in germ-free mice), gnotobiotic mice (defined flora present), or specific pathogen free mice (part barrier, to keep mice free of a known range of microorganisms).

Cages may be of stainless steel or plastic, and are usually of the shoebox type. Bedding should be provided, such as wood chips or ground corn cobs. This should be autoclaved, or at least screened to ensure it is uncontaminated before use.

Adult males tend to fight, and are usually housed alone. Female mice are less aggressive, and can be kept in groups of familiar females. Females with litters will defend their young and are best separated while nursing.

Keeping mice in compatible groups, and placing cardboard rolls and objects to climb on in the cage all enrich the environment and reduce stereotypic behaviour such as barbering.

Feeding

Mice are usually fed *ad libitum* with a complete pelleted mouse diet, from hoppers suspended above the floor to prevent faecal contamination. These should be cleaned twice weekly. Generally **mice will consume 3–5 g of pelleted diet daily**, but there are strain differences, and disease states and pregnancy affect the food requirements.

Water

Water is required for lubrication of the food as well as hydration. It is best supplied in automated systems or bottles, not open bowls. **The usual requirement is 6–7 ml water daily.**

The water may need to be acidified, chlorinated or sterilized to reduce contamination, particularly for immunocompromised mice.

When ill, mice drink very little and rapidly dehydrate. Medicines administered in the water are therefore unlikely to be effective, and care must be taken to ensure that adequate quantities of fluid are administered by other means (see Chapter 13).

Environment

Mice have a large surface area to volume ratio, and therefore lose heat rapidly and are sensitive to changes in ambient temperature. Much energy is expended in maintaining body temperature, and they cannot tolerate a reduction in room temperature, which will occur if heating levels are reduced to conserve energy at night.

Mice are also susceptible to water loss. They cannot afford to sweat or pant to lose heat, as this would cause dehydration. In the wild they use behavioural mechanisms such as burrowing to keep cool, which cannot be done in most laboratory housing. Therefore maintenance of the correct environmental conditions is vital.

To comply with the Home Office Code of Practice, mice require temperatures between 19°C and 23°C, humidity of 40–70 per cent, 12–15 air changes per hour, and 12 hours daylight daily. According to the Home Office Code of Practice, the light intensity should be 350–400 lux at bench level. For albino mice, it should be less than 60 lux in the cage to avoid damage to the retina.

Mice are also very sensitive to ultrasound. Normal noise levels may sound quiet to the human ear, but they may be extremely loud for a mouse. Care should be taken to reduce ultrasound in rodent-keeping facilities.

Breeding

Male mice reach puberty at 7 weeks, and females at 6 weeks. Females then cycle every 4–5 days. This is effected by photoperiod and the presence of others. Oestrus, mating, and ovulation tend to occur during the dark phase of the light cycle. Group housed females will become anoestrous, but will resume cycling if a male is introduced. This is known as the Whitten effect, and can be used for synchronizing oestrus or for timed matings.

Mating results in the formation of a vaginal plug, which can be detected to confirm mating. Gestation lasts 19–21 days, and 1–12 pups may be born, depending on the strain. They are weaned at 3 weeks.

Mice may be bred using a harem system, with one male for 2–6 females, pregnant females being removed from the group to give birth, or may be kept together in a monogamous system. With the latter system, the young are removed before the next litter is born.

Mice will breed until they are 12–18 months old.

Growth

Strains vary dramatically in rate of growth. In general, outbred strains grow much faster than inbred strains. (See Figure 6.3).

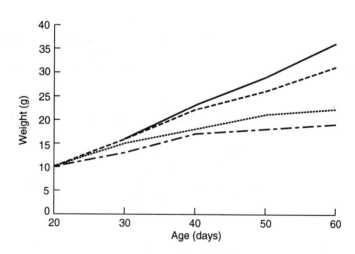

6.3 Growth chart for typical outbred albino and inbred albino (BALB/c) mice. ——, outbred albino male; ------, outbred albino female; ··········, BALB/c male; ––·–·, BALB/c female.

Rat

Rats used in research are mainly derived from the brown or Norwegian rat, *Rattus norvegicus*. Outbred and inbred strains are available. Commonly used outbred strains include the Wistar and Sprague Dawley varieties. Fewer inbred strains exist than for mice, but a commonly encountered one is the Lewis rat.

Behaviour

Rats are usually friendly and amenable animals if handled gently, although there are some strain differences. They will become more friendly with more frequent handling. Although they will live in single cages happily, they can be social and will live together in groups with little fighting provided they are not overcrowded (see Table 6.1 for Home Office Guidelines).

Rats are naturally curious, and will explore in any new situations. Failure to stand erect and take an interest in their surroundings is an indicator of poor health (see Chapter 11).

Rats have a tendency to be nocturnal. Feeding, drinking, and mating all tend to occur at night. Their eyesight is poor, and blind rats will behave as if perfectly normal. The Harderian gland, a modified tear gland situated on the bulbar conjunctiva of the third eyelid, produces a porphyrin-rich secretion which normally lubricates the eye. When the rat is stressed, this secretion tends to overflow onto the face, producing a red ring around the eye which is characteristic of stress. This is known as chromodacryorrhoea.

Housing

Rats may be kept in stainless steel or plastic cages. If mesh floors are used, care must be taken to ensure that the mesh is small enough that young animals do not fall through it, but not so small that the rats catch their feet. Mesh floors are not suitable for breeding females, as nest building is not possible. Solid bottomed cages are best, and woodshavings, wood chips or paper may be used as bedding. The bedding should be changed two or three times weekly, and the cage sterilized once or twice weekly, or they will become very dirty.

Rats like to stand erect, and so cages with high lids are required (see Table 6.1 for minimum height recommendations).

Rats can be kept in conventional units, or with full or partial barriers to produce animals with no, or limited, pathogens.

Feeding

Rats, like all rodents, are coprophagic. They can be fed *ad libitum* on a complete pelleted rodent diet, from hoppers suspended above the floor of the cage. Food hoppers should be cleaned once or twice weekly. The diet should contain 20–27 per cent protein. Higher protein levels than this may reduce reproduction efficiency. Rats are cautious eaters and will reject strange food.

Rats will eat 5 g feed per 100 g bodyweight daily.

Water
Water may be provided by sipper tubes or by automated watering systems. The system should be cleaned once or twice weekly. The water may need to be acidified, chlorinated or sterilized to reduce contamination, particularly for immunocompromised rats.

Rats will drink 10 ml water per 100 g bodyweight daily.

Environment
Rats are less sensitive to temperature changes than mice, but prefer to be kept between 19°C and 23° C. Young rats have much brown fat to assist in thermogenesis, the level of which reduces with age.

The humidity should be 40–70 per cent. Low humidity results in ring-tail, in which an annular lesion appears around the tail, which may result in sloughing of the tail distal to the lesion.

A 12 hour light period is adequate for rats but, being nocturnal, bright light is deleterious particularly for albino rats and results in retinal degeneration. The level should be less than 400 lux, or 100 lux for albinos. Photoperiod affects the oestrous cycle, and 12 to 16 hours light is best for optimal breeding.

Ventilation is particularly important for rats, as many of their pathogens are aerosol-borne. Twelve to fifteen air changes per hour is sufficient, provided the air is not recycled or an effective filter is present.

Breeding
Puberty occurs at 50–60 days, and breeding begins at 3 months, when females weigh 250 g and males 300 g. They breed until they are 12–18 months old.

Oestrus occurs every 4–5 days. The Whitten effect is less pronounced in rats than mice, (synchronization of oestrus in females by exposure to male pheromones), but does occur. Mating usually occurs at night, and a copulatory plug of gelatinous material is left in the vagina for 12–24 hours, which then falls out and can be detected to confirm that mating has occurred. Gestation lasts 21–23 days. A litter of 6–12 pups is born in a shallow nest made by the female. Paper, wood shavings or specialized bedding materials (e.g. 'Vetbed' simulated fur bedding) can be supplied to aid in nest building, but are not essential. Ground corn cobs are not suitable for nest-making. Weaning occurs at 21 days.

If a female is disturbed in the immediate post-partum period, she may destroy her young, so extreme care must be taken when cleaning cages during this time.

Rats may be bred by monogamous or polygamous mating systems.

With a monogamous system, the female will be mated at the post-partum oestrus, and the young are removed at weaning. This produces the maximum number of litters, but the male may interfere with the young. He can be removed at parturition, and returned to the female after the young are weaned. If the female is lactating during gestation, implantation can be delayed, leading to a 3–7 day increase in the length of gestation.

A variation in the monogamous system exists in which a single male is moved between singly housed females, spending a week with each. One male is used for every seven females. Care must be taken to remove the male before parturition, or he may attack the litter.

In polygamous systems, one male is housed with 2–6 females. Pregnant females are removed prior to parturition and returned after weaning. Females in this system produce more milk and have larger litters.

Growth
Male rats exhibit prolonged growth, and bones do not become fully ossified until their second year.

Inbred and outbred rats differ slightly in their rats of growth. (See Figure 6.4.)

Guinea-pig
The guinea-pig, *Cavia porcellus*, is a rodent. There are several different varieties of guinea-pig available, including the short-hair

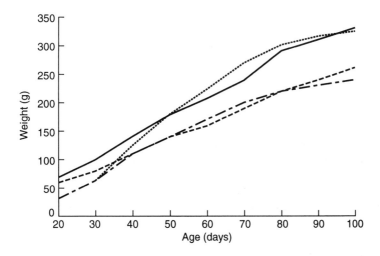

6.4 Growth curves for inbred and outbred rats. ———, outbred male; ------, outbred female; ············, Lewis male; –·–·–, Lewis female.

(English and American varieties), Abyssinian (which have hair in whorls), and Peruvian (which have long hair).

Commonly used laboratory strains are derived from the short-hair variety. The Dunkin–Hartley and Hartley guinea-pigs are outbred strains, and strains 2 and 13 are inbred strains.

Behaviour

Guinea-pigs are amenable animals which rarely bite. Naturally, they are crepuscular but in the laboratory they will be active for periods throughout the day and night.

When startled, guinea-pigs have a tendency either to become immobile or to stampede and vocalize. In groups, this latter instinct may result in trampling of the young. Barriers within the cage can reduce this problem.

The approach of a person will cause excitement, and the scatter reaction should be elicited as an attempt is made to capture a guinea-pig. The normal behaviour is for the guinea-pig to 'resist arrest' and vocalize strongly. If this does not occur it may indicate that there is a problem.

Group housed familiar guinea-pigs will soon establish a stable hierarchy, which is male-dominated and maintained mainly by olfactory cues, but with some barbering and chewing of subordinate males. If unfamiliar males are placed together fighting will ensue particularly in cramped conditions or if oestrous females are present.

Guinea-pigs are creatures of habit and become increasingly unable to cope with changes in routine as maturity approaches. If there are any changes in the type of food hopper or water bottle, or in the type of food or water, the guinea-pig may be unable to adapt and cease eating and drinking. This is particularly disastrous with pregnant females. Similarly, if there are changes in the type of housing, problems may be encountered. Guinea-pigs reared on mesh may cope well with it, but inexperienced animals can fall through it and may break their legs.

Housing

As gregarious animals, guinea-pigs like to be housed in groups. This may be in floor pens (See Figure 6.5), or large plastic or steel cages (for dimensions see Table 6.1).

Although guinea-pigs rarely jump, cages should have sides at least 23 cm high, and more height is required for open-topped floor pens.

Guinea-pigs thrive on solid floors, but may be kept on slats or mesh if accustomed to these types of floor, otherwise mesh floors can produce problems as described above. Mesh floors may also predispose to foot pad ulcers and increased stress levels, and are contraindicated with experiments involving joints and feet.

Bedding may be provided in the form of wood shavings, shredded paper, or sawdust together with hay. Fine shavings and sawdust alone may cling to moist areas such as the perineum and probably are best not given to breeding guinea-pigs. Larger shavings are preferable for these animals.

Guinea-pigs are messy animals and will disperse opaque, creamy coloured urine, and faecal pellets throughout the pen. All pens, cages, feeding receptacles, and water bottles must be cleaned and disinfected at least 2–3 times weekly. Removal of urine scale may require the use of acidic cleaning agents.

Feeding and watering

As guinea-pigs are messy, food and water bowls placed on the floor will soon become soiled with bedding, urine, and faeces, and should be suspended above the floor or cleaned frequently. There is a tendency to play with drinkers, which leads to messy floors, and bottles quickly become empty. Automated watering systems ensure a constant water supply, but in solid floored systems, care must be taken to prevent flooding. All watering systems need to be checked and cleaned frequently. Any changes in watering system will upset the routine, and the guinea-pig will need help to adapt.

The water requirement of a guinea pig is 10–40 ml per 100 g bodyweight daily.

Guinea-pigs are fastidious eaters and will reject unfamiliar food. They require a pelleted, freshly milled complete guinea-pig diet, not one designed for any other species. Supplements of hay or greens

6.5 Guinea-pigs housed in floor pens.

may be given, but with care as digestive disturbances may result from an excessive amount, or from greens which have not been properly washed.

The food requirement is 6 g per 100 g bodyweight daily. However, much food is wasted and more should be supplied.

The food should contain 18–20 per cent crude protein, and 10–16 per cent fibre. Guinea-pigs are unable to synthesize vitamin C, and require 5 mg/kg daily normally, and up to 30 mg/kg if pregnant. This can be supplied in the food or water, or by giving cabbage, kale or oranges. Food with added vitamin C must be used within 90 days of manufacture, or the vitamin C will degrade.

Coprophagy does occur in the guinea-pig, but may not be essential.

Environment
Guinea-pigs thrive at temperatures between 18°C and 26° C, with a humidity of 40–70 per cent. They should have 12–15 air changes per hour, and 12–15 hours light daily.

Breeding
Female guinea-pigs reach puberty from 5 to 6 weeks, and males from 8 weeks. The average is 9–10 weeks. Pairing should be done when the female is 400 g (at 2–3 months), and the male 650 g (3–4 months). One boar can be housed with 1 to 10 females.

The oestrous cycle of the female lasts 15–17 days, and she is receptive for 6–11 hours. The vagina is covered by an epithelial membrance which is intact except during oestrus and parturition, both of which are signalled by perforation of the membrane.

Gestation lasts 59–72 days, depending on litter size. In the last week of gestation, the pubic symphysis separates under the influence of the hormone relaxin, and once the gap reaches 15 mm parturition will take place within 48 hours. Females should have their first litter before reaching 7–8 months of age, or the symphysis will be unable to separate sufficiently and dystocia will result. In any case, there is often a high incidence of dystocia and fetal death. Abortions and stillbirths are common.

Female guinea-pigs can breed until they are 20 months old. Thereafter, the litter size tends to drop and dystocia is more common.

Neonatal guinea-pigs are precocious, and weigh 60–100 g. They can eat solid food within a few days. Hand-rearing is not difficult making Caesarean rederivation of colonies quite easy. The young are not hungry until 12–24 hours after birth, and can then be fed cows' milk or soaked guinea-pig pellets. If the females are not kept in harem groups, the young may be removed at birth and hand-reared, to allow the sow to be mated at the post-partum oestrus. Otherwise, weaning takes place at 180 g (15–28 days), or 21 days (165–240 g). Weaned males intended for breeding need to be weaned late or

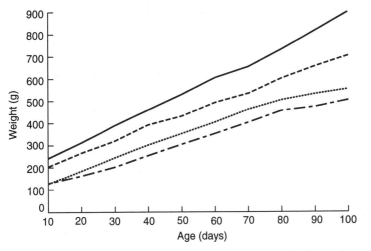

6.6 Typical guinea-pig growth chart. ———, Hartley male; ------,
Hartley female; ⋯⋯⋯, average male; —·—·, average female.

group housed to allow development of normal adult reproductive
behaviour.

Growth
The growth depends largely on the strain of guinea-pig (See Figure 6.6). Young guinea-pigs should gain 2.5 to 3.5 g daily up to
60 days.

Gerbil
Over 100 species are included in this group, but the one used
mainly for research is the Mongolian gerbil, *Meriones unguiculatus*.
These are easy to keep, being hardy with few diseases. Most gerbils
used are from outbred stocks, but an inbred strain is available
(MON/Tum strain).

Behaviour
Gerbils are generally docile creatures, which are easily handled and
rarely bite. They are generally very active, and when approached,
they will resist being caught. Normally they exhibit exploratory
behaviour in new surroundings, and if loose they do not hide but
show curiosity and interest in the environment. In the wild they are
crepuscular.

Although friendly with people, unfamiliar adults caged together
will be aggressive. Stable groups may be established by putting
animals together before weaning. Normal social behaviour will then
be seen, in which animals wrestle and groom each other.

Some young gerbils will display epileptiform seizures if stimulated, for example by handling or environmental disturbances. Frequent handling from an early age will reduce the frequency of seizures. Leaving the gerbil in a warm, dark, quiet place to recover is the best treatment for a seizure. Drugs may be given to control a severe seizure, but are generally contraindicated as they can cause death.

Housing

Gerbils have similar husbandry needs to other rodents. They prefer solid floors to mesh, and need at least 2 cm depth of bedding for nest building, which occurs even if the female is not pregnant. Sawdust or shavings made from pine should not be used, as the fur tends to become matted with these materials.

Gerbils need at least 15 cm space between the top of the bedding and the roof of the cage, as they like to sit erect. As gerbils tend to gnaw the cage, it should be of suitably strong material to resist attempts to escape.

Very little urine is produced by gerbils, and their faecal pellets are small and hard. They are naturally clean animals, and cages usually need cleaning at weekly intervals only.

Feeding

Eating is spread throughout the day and night and like all rodents, gerbils are coprophagic. Standard rodent diets with 22 per cent protein are adequate, but due to an unique lipid metabolism the dietary fat level must be below 4 per cent to prevent high blood cholesterol developing. Higher fat levels lead to obesity, and in females may cause infertility due to fat deposition around the genital tract. Standard food hoppers are normally used, but supplementary food may be put on to the floor for young gerbils until they become used to the hoppers. Pellets can also be soaked.

Gerbils consume 5–8 g of pelleted food daily.

Water

Gerbils, being desert animals, will cope with very little water. They produce very concentrated urine, and are resistant to water loss. Older males tend to need more than younger animals. Bottles or automated systems may be used, and care must be taken that the sipper tube is accessible to all individuals in the cage, including juveniles. **Gerbils do well with 4–10 ml water daily.**

Environment

Due to their desert origin, gerbils can tolerate high temperatures and have a great ability to regulate their body temperature. They thrive between 19° C and 23° C. Humidity should be between 30 per cent and 50 per cent, and 12 hours light should be provided daily at 350–400 lux. Ventilation should produce 12–15 air changes per hour.

Breeding

Gerbils are less efficient breeders than other rodents. Often, when mates are introduced fighting occurs, or if a long-term partner dies, the remaining partner will not breed again. Less aggression is seen if the introduction takes place on neutral territory.

The best regime involves keeping gerbils in monogamous pairs and never separating them. If they are separated then reintroduced, they will probably fight.

Puberty occurs from 6 weeks of age, and they are usually bred from 9 to 12 weeks. Oestrus occurs in the female every 4–6 days. Gestation lasts 24–26 days, unless the female is mated at the post-partum oestrus. In this case, lactation prolongs the gestation period to 27–48 days. To avoid post-partum mating, the male can be removed at parturition, but the separation should be for less than two weeks.

Four to five pups are born, and the male will assist in caring for them. Despite this, neonatal mortality is high. Weaning takes place at 21 days.

Growth

Male and female gerbils show a similar pattern of growth.

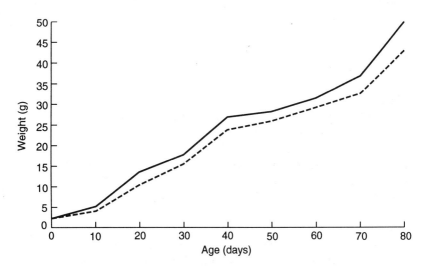

6.7 Typical gerbil growth chart. ———, male; -------, female.

Hamster

Hamsters are known widely for their cheek pouches, in which they gather food, to store in deep burrows, particularly when photoperiod is decreasing. The pouches are unusual in having no lymph drainage, and therefore do not reject tissues transplanted to them. This is useful in immunological and other research.

Several different varieties of hamster are used, the most common one being the golden or Syrian hamster, *Mesocricetus auratus*. Different strains are available.

Behaviour

Hamsters are readily tamed, and rarely bite unless startled or handled roughly. Males are more docile than females. They are active at times during the day, but are mainly nocturnal, and most activity occurs at night.

Hamsters are solitary animals, and will attack each other. Females will attack males except for a brief period during oestrus, and often attack other females. Groups of same sex animals may be maintained if they are put together at weaning or before puberty.

Territory marking is done using a secretion from the flank glands, which are visible, particularly in the male, as dark patches on each side.

Housing

Hamsters, being solitary, prefer to be housed individually. They are adept at chewing through cages, so tough plastic cages with solid bottoms are used. If they do manage to escape, unlike rats and mice, they will not return to their cages. Deep piled bedding is usually provided. Little waste is produced, and cages can be cleaned out 1–2 times weekly.

A breeding female must have deep bedding and soft nesting material such as paper or Vetbed simulated fur bedding, or an inadequate nest will lead to her abandoning or killing her young. To avoid disturbing a nursing mother, cleaning out may be done every 2 weeks.

Feeding

The requirements of hamsters have not been specifically studied. They are coprophagic, but may have different digestive systems to rats or mice. A diet with 16 per cent protein, 5–7 per cent fat and 60–65 per cent carbohydrate would appear to be sufficient. **Hamsters eat 5–7 g pelleted diet daily.**

As they have blunt noses, hamsters cannot feed from standard wire mesh suspended hoppers. They need hoppers with slots greater than 11 mm, so they can pull the food through on to the floor. Nursing females should be floor fed anyway, as they become preoccupied with pulling food from the hopper and neglect their young.

Food hoppers and drinking apparatus should be cleaned once or twice weekly.

Water
Hamsters need 10 ml water per 100 g bodyweight daily. Water bottles or automated systems can be used, but the sipper tubes must be stainless steel and not glass, as hamsters can bite through glass.

For breeding animals, the sipper tubes should extend low into the cage so that neonates can reach them. Lactating females have a greater water requirement.

Environment
Hamsters are originally from hot countries, and burrow to avoid heat. If unable to burrow, they tolerate heat poorly. Cold however may be tolerated quite well. If the temperature drops below 5° C, they can go into a state of pseudo-hibernation, from which they can be aroused by stimulation. If insufficient food is available, hibernation is delayed.

Hamsters should be kept between 19°C and 23°C, and breeding animals need to be at the higher temperature. Relative humidity should be 40–60 per cent. A 12–14 hour period of light should be provided daily.

Breeding
Puberty occurs at 32–42 days, and breeding usually starts at 6–10 weeks for females, and 10–14 weeks for males, when they weigh 90–130 g. Fertility is lower in winter. The female has oestrus every 4 days, and is receptive for a very short time. Gestation lasts 15.5–16 days. Between 4 and 12 young are born. Due to the short gestation, they are small, weighing 2–3 g, and immature. An adequate nest is essential to provide warmth and security for the young, hence the need for nesting material. Weaning occurs at 20–25 days.

Males and females are usually kept separately, and put together for a short period after dark for mating. If the female is not receptive and does not accept the male, he should be removed at once. Otherwise, he is removed at the end of the dark period, before the light part of the cycle begins.

A system of monogamous pairs can be maintained if the male and female are put together before puberty and kept permanently together. Alternatively, females can be rotated through the male's cage at weekly intervals, one male being used for seven females, or a harem can be set up with several males and females together. The females are removed prior to parturition and returned after weaning. However, these systems can lead to fighting.

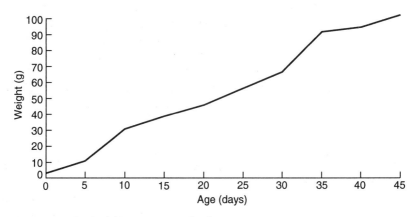

6.8 Typical hamster growth chart.

Hamsters commonly abandon or kill their young. This can be triggered by environmental disturbances, inadequate nesting material, or early handling of mother and young. Sometimes when disturbed, females hide their young in their cheek pouches. While this affords the young some protection, it may also suffocate them.

To avoid stressing a nursing mother, sufficient food and bedding for at least one week should be provided just prior to parturition, so that she does not have to be disturbed for 7 days post-partum.

Fostering and hand-rearing of young are usually unsuccessful, so it is difficult to rederive a colony by Caesarean section to control disease outbreaks.

Growth
The hamster has a growth spurt between four and five weeks of age.

Rabbit
Laboratory rabbits are derived from the domestic rabbit, *Oryctolagus cuniculus*. They are well known for being gregarious, and for their burrowing tendency.

Many strains are available, the most common one being the New Zealand white, a large outbred rabbit. The smaller Dutch rabbit is also popular. Inbred strains are available.

Behaviour
In the wild, rabbits are nocturnal or crepuscular, emerging from their burrows to feed at night. In the laboratory, activity may be seen day and night.

Rabbits are amenable animals which can readily be trained, and which rarely bite. If handled gently, rabbits are docile. If handled roughly, however, they will kick and scratch with their powerful hind legs. If poorly supported during struggling, contractions of the strong spinal muscles can cause damage to the vertebrae and spinal cord.

Natural aggressive behaviour may be seen in the laboratory in breeding rabbits and in those undergoing puberty (at 3 months of age). Fighting, thumping, and burrowing are the usual manifestations of aggression. Breeding animals are therefore best kept apart.

Housing

Rabbits require strong cages with no rough or sharp edges. The floor may be of mesh, or solid. Rabbits do particularly well in floor pens, with high sides. Groups of young females or does with litters can be kept very successfully in such pens, with deep bedding and play objects such as cardboard boxes. Generally, male rabbits are separated at weaning and females at puberty to reduce fighting. Females housed together may also have low fertility. These potential problems of group housing are reduced by the provision of ample space and by adding 'bolt holes' within the pen where rabbits may escape (Figure 6.9).

Rabbit housing should be cleaned at least weekly. They produce copious, turbid urine, which may be yellow to dark red, due to the presence of a varying quantity of porphyrins. The urine tends to leave scale on the litter trays due to the calcium content, and so they may need to be cleaned with acidic agents. Absorbent tray liners help to lock in ammonia and to reduce the scale problem.

6.9 Floor housed rabbits with examples of environmental enrichment.

Feeding

Rabbits are coprophagic, and this is an important part of their digestion and nutrition. They eat pelleted diets, and require a diet with a high fibre content. Much fibre remains undigested, but it is required for bulk. A diet with 12–22 per cent fibre, and 12 per cent protein for maintenance or 15–17 per cent for growth allows *ad libitum* feeding but does not produce obesity.

A high fibre diet also reduces the incidence of hairballs and diarrhoea. Hay may be given as a supplement to reduce the formation of hairballs.

High energy diets are required for rabbits which are reproducing and for the dwarf breeds, with 10 500 kJ/kg feed. For maintenance, 8800 kJ/kg is sufficient.

Rabbits need 5 g of high energy food per 100 g bodyweight daily.

As gut flora play an important part in digestion, changes in diet should be done gradually, over a 4–5 day period, to allow the flora to adapt. Failure to do this will result in diarrhoea or anorexia.

Water

Water should be supplied *ad libitum*, and be fresh and clean. Automatic systems are often used.

Rabbits normally consume 10 ml water per 100 g bodyweight.
Lactating does may drink up to 90 ml per 100 g bodyweight.

Environment

Rabbits require temperatures between 16°C and 20°C. Neonates cannot maintain their body temperatures until they are 7 days old, so they must be kept in a warm environment. Humidity should be kept between 40 per cent and 60 per cent.

Females require 14–16 hours of light daily, and males 8–10 hours. Shorter light cycles may result in reduced sexual activity in the autumn.

Ventilation is particularly important for rabbits. They are susceptible to respiratory diseases, and poor ventilation allows a build up of ammonia which predisposes to these. Draughtless ventilation and efficient tray liners reduce the ammonia level. At least 12–15 air changes per hour should be provided.

Breeding

Females begin breeding at 4.5–7 months, males from 6 to 7 months. Rabbits are induced ovulators, and have no oestrous cycle as such. They are receptive for 7–10 days, then inactive for 1–2 days. Does are also receptive at intervals during pregnancy and lactation. There is some seasonal effect, but if day length is maintained, and particularly if the temperature is high, breeding will take place all year. Breeding efficiency falls after 7–11 litters, or 3–4 years.

The doe should be taken to the buck, and removed if mating does not occur within a few minutes. Coitus induces ovulation and results in pregnancy in 75 per cent of does. Pregnancy lasts 31–32 days. Care should be taken in handling the doe during gestation, as the pregnancy is easily aborted.

The doe must have a clean nest box in the last week of gestation. The cage may have a solid floor together with a raised wire floor on which there is plenty of bedding. This construction keeps the nest dry and clean. A doe will scatter her young if the nest is dirty, or if it smells of disinfectant. The doe will line her nest with hair. A cardboard box placed box on its side makes a cheap and disposable nest box.

Does usually kitten in the morning. Between 1–22 kits may be born, usually 7–8. This varies with the breed. It is rare for them to be eaten by the doe unless there is a deformity or they are dead, but it may also happen if the doe is inexperienced or if there is some environmental disturbance.

Does will rarely retrieve their young if they crawl out of the nest, hence the need for a good nest box. Young are suckled only once daily, in the morning, and lactation lasts 6–8 weeks. Hand-rearing of rabbits is relatively easy.

Does may be rebred shortly after parturition, but more usually are mated after weaning.

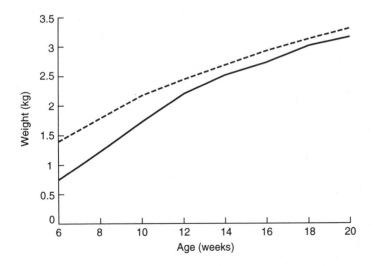

6.10 Typical New Zealand white rabbit growth chart. ——, minimum weight; -------, maximum weight.

Growth

Growth varies with the breed of rabbit. See Figure 6.10 for the growth curve of the New Zealand white rabbit.

CARNIVORES

Dogs, cats, and ferrets are the carnivores used most frequently in the laboratory. They have less stringent environmental needs than the rodents and rabbits, and are familiar to most people, particularly cats and dogs.

Dog

The dog, *Canis familiaris*, has been domesticated for thousands of years. Many breeds are recognized and purpose-bred laboratory dogs are usually beagles. Dogs are useful models for many human diseases, and because of their size and good temperament they are particularly suited to studies requiring close monitoring.

Behaviour

Dogs are gregarious animals, and in the wild live together in packs with males and females having dominance hierarchies. Members of the pack co-operate in hunting, etc. Dogs should be kept in groups or in pairs in the laboratory if possible, with care being taken to ensure that subordinate members of the group get sufficient food.

Dogs need to be socialized to man and other dogs as puppies, between 6 and 8 weeks of age, or they will be unapproachable by 14 weeks. Social contact is therefore vital, and must be maintained if the dog is to remain friendly. Well-socialized dogs are much easier to handle and make better experimental subjects than unsocialized dogs, as the former are much less stressed by handling and procedures. Puppies should be allowed to mix and play together from a young age. They should also be trained to accept being in a subordinate role, with the handler dominant. Dogs are intelligent and learn readily, which can be useful in the laboratory. Male dogs are generally more aggressive than females, but this varies with the breed. Beagles tend to be placid, amenable dogs which adds to their attraction.

Dogs have many methods of communication, with other dogs and people, with olfactory signals being important. Males mark territory by cocking their legs, and frequent cleaning of the pen will result in frequent urination to re-mark the territory. A balance has to be struck between the requirements of the people and those of the dogs. Dogs can be encouraged to urinate in particular places by the selective use of odour eliminators. These will inhibit territory marking if used in areas which are required to be clean, such as the bed area and feeding area.

Housing

Dogs can be kept in indoor pens, with or without outside runs. If the runs are large enough, there will be no soiling of the bedding area. There must be facilities for adequate exercise, either in the individual runs or in a communal exercise yard where dogs can go in pairs or groups. Cages are unsuitable for dogs long term as there is little room for exercise or human contact.

There must be a warm dry area for the dog to sleep in. Beds may be provided, which should be resistant to chewing, and may be raised above floor level. There may be underfloor or overbed heating. Sawdust litter may be provided in the bedding area.

The floors and walls need to be of smooth, impervious material which can withstand frequent cleaning. Faecal material should be removed from all areas daily, and the pen thoroughly washed once or twice weekly. The exercise run should be cleaned more frequently. Puppies are not as clean as adults, and pens for nursing bitches need to be cleaned two to three times daily.

Feeding

Commercial tinned and dry diets are available. Tinned foods are usually given with a cereal-based biscuit. Dry or semi-moist diets are complete, and just need to be fed with water. They are usually fed *ad libitum*, so the dog adopts a little-and-often pattern of feeding. If there is a tendency to obesity, this may be restricted. Feeding infrequent, large meals of dry food predisposes to acute gastric dilatation and should be avoided. Sudden changes in diet are not accepted well and may lead to digestive disturbances. Cold food is less well tolerated than warm food.

Diets usually contain 22 per cent protein and 5 per cent fat. Dogs are very adaptable and deficiencies are rare.

A 13 kg beagle needs 0.8 kg canned food or 0.25 kg dry food daily for maintenance.

The ration should be increased by 30 per cent for the last 3 weeks of gestation, and raised to 3 or 4 times maintenance during lactation. For peak lactation, a high-energy food with a high calcium level should be fed.

Water

The water requirement depends on the diet and the size of the dog. An unrestricted supply is usually given.

The total water requirement is 70–80 ml/kg/day. This is provided by food and water. An adult beagle needs approximately one litre daily.

Environment

Dogs are very adaptable. They can cope with temperatures between 15°C and 24°C, and lower temperatures are tolerated if there are no draughts and the dogs are in groups. Extremes should be avoided. Neonatal dogs require 30–32°C: some of this is provided by the bitch but the air should be at least 26–28°C for the first 5 days. By 4 weeks, 24°C is adequate.

Ventilation providing 8–12 air changes per hour is sufficient, and natural daylight is preferred.

Breeding

Puberty is reached at 7–8 months in the male and 8–14 months in the female. Bitches are monoestrous: each cycle is followed by a period of anoestrus lasting 4–14 months. The cycle starts with pro-oestrus, in which there is a sanguinous discharge. The discharge reduces as the bitch enters oestrus, after 6–10 days. The bitch is receptive for 6–12 days, then enters metoestrus, when activity declines. The bitch only permits mounting during oestrus, and the male and female will 'tie' for 20–30 minutes during copulation. Ovulation is spontaneous and pseudo-pregnancy is common in the bitch. Gestation lasts 59–67 days. An average of six puppies is born, at intervals of up to 1 hour. Puppies must get colostrum during the first 24 hours, and are fed initially every 2–4 hours. The bitch licks the puppies to stimulate respiration at first, then regularly licks the perineal area of each puppy to stimulate urination and defaecation for the first 3 weeks.

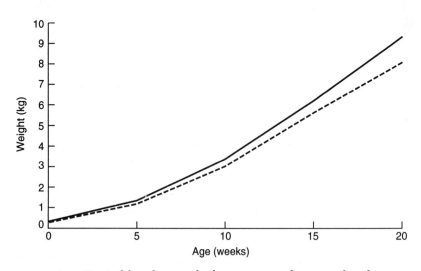

6.11 Typical beagle growth chart. ———, male; -------, female.

Puppies are weaned at 6 weeks, when they can eat solids. Growing puppies need to eat twice the maintenance level of an adult of similar size.

To breed dogs in packs, one male is put with up to 12 bitches, and bitches are removed from the pack for whelping. Alternatively, bitches can be kept in pairs and moved in with the male 10 days after the onset of pro-oestrus for five days.

Growth
Puppies should multiply their birthweight by a factor of 40 or 50 in their first year (see Figure 6.11). Adult weight is reached at 9 months. Before this, they must be fed for maintenance and growth.

Cat
The domestic cat, *Felis catus*, has possibly been derived from the wildcat, *Felis sylvestris libyca*. Cats are true carnivores and hunters, using their good sight and hearing to find prey. The reflective tapetum lucidum in the eye allows for particularly good night vision.

Several breeds of cat are available, but purpose-bred non-pedigree cats are usually used in the laboratory.

Behaviour
Cats are solitary animals with the capacity to be sociable. Feral cats are nocturnal, and mark territories with urine and secretions from their anal glands. Males defend large territories covering the smaller territories of several females, and are polygamous.

Cats communicate by vocalization, facial expression, postural changes, and scent. They will demonstrate affection and aggression towards other cats and people, but if unused to people will be nervous if approached. They are sensitive to the attitudes of staff, and a gentle caring approach will soon result in friendly cats.

Young cats are playful and agile, and are kept in groups in large cages with many playthings. Adult females can be put together and will be amicable after the first few days. Adult males, however, will fight over females, so only the dominant males mate. Males which are not littermates are usually separated at 4-6 months.

Cats spend 60 per cent of their time asleep, and the remainder of the time is spent patrolling and marking territories.

Housing
Cats may be kept indoors or outdoors. Group housing is best, with single caging for adult males, pregnant females, and queens with litters for the first 4–6 weeks. Singly housed cats should be allowed to exercise daily, and ideally should be able to see other cats. In a breeding group, a large area is needed with perches and refuges,

where kittens can play and where the cats can escape to, for example during cleaning, when they like to be above the floor. Perches should be sterilized regularly.

All cats require scratching posts and sleeping boxes, which may be hung from the wall to facilitate cleaning. The floor is usually solid, but grids are sometimes used. A dirt tray must be provided away from the sleeping area for urination and defaecation. These must be cleaned out at least once daily, and the rest of the room weekly.

Feeding

As carnivores, cats have a very high protein requirement, and specific requirements for particular amino acids (taurine and arginine). **Protein must constitute at least 30 per cent of the diet on a dry matter basis.** Cats also become bored easily, and variety is preferred. Many commercial diets are available, tinned, dry, and semi-moist. Food must be fed fresh with no rancid fat, or it will be rejected. Twice daily feeding of tinned food with *ad libitum* dry food overnight allows little and often feeding with no deterioration of the food. Food bowls may be placed directly on the floor or suspended from the wall, but must be able to be cleaned daily to remove stale food.

Water

On a dry diet, **cats will drink 200–300 ml daily**, and much less on tinned diets. Cats dislike drinking from nozzles, so automatically filling bowls are best.

Environment

In outdoor units, the environment is not controlled. Cats cope well but have litters mainly in spring and summer. In closed units the photoperiod can be controlled, and with a 12 hour light cycle litters are spread throughout the year. The temperature needs to be between 21°C and 25°C, with humidity between 45 per cent and 65 per cent. Ventilation must be draught-free, and 10 to 15 air changes per hour is sufficient. Filters must be placed over the vents to prevent them from becoming blocked with hair.

Background noise helps keep cats calm. In very quiet environments cats will startle easily.

Breeding

Puberty is reached at 6–7 months, but breeding does not usually take place until 1 year. Outdoor cats breed seasonally, with females receptive from winter to summer. Indoor cats breed all year. Females are polyoestrous, with cycles lasting 14 days. Oestrus is signalled by vocalization (calling), and postural changes (lordosis). Coitus induces ovulation, and it is normal for several matings to be required before ovulation occurs. The queen should be taken to the tom for mating.

Pregnant females should be removed from the colony from 10 days before birth, but should be allowed to mix with the other females for short periods in the morning and afternoon so as not to lose their places in the hierarchy. Gestation lasts 63 to 67 days. There may be between 1 and 10 kittens in the litter (average 4). Kittens need highly concentrated protein-rich milk for up to 7 weeks, but can be weaned as early as 4 weeks. Once kittens are 6 weeks old, they and their mother can return to the group.

Growth
Kittens are born weighing 90–140g, and should gain 80–100g weekly.

Ferret
The ferret, *Mustela putorius furo*, is important as a laboratory animal as it represents the Carnivora, but is small enough to be kept easily in the laboratory. There are two main varieties, the fitch ferret, which is buff with a black mask and points, and the albino.

Behaviour
Ferrets are usually friendly, particularly if handled during rearing. They are very curious, and like to explore and burrow. In strange surroundings, they can be nervous and may bite, particularly if roughly handled. Females with litters are protective and may also bite.

Generally, unless breeding, ferrets like to be group housed, and young ferrets will readily play together. Ferrets communicate by using musk glands, which are situated lateral to the anus. The secretion may also be expressed when excited or frightened, or during the breeding season.

Housing
Ferrets are gregarious. Females can be housed together except when they have kits, and males only need to be separated in the breeding season. Standard rabbit cages are often used, but ferrets prefer solid floors with bedding such as sawdust or shavings. Urination and defaecation tend to occur in one place so cleaning out is only necessary once or twice weekly, and ferrets can be trained to use a litter box.

Nest boxes are usually provided for security and warmth for females with litters, and plastic tubes, boxes, and paper bags will add to the richness of the environment.

For females with litters, a solid partition is required at the bottom of the cage door to prevent the young from falling out.

Feeding

The nutritional requirements of ferrets have not been extensively studied. They cope well on mink diet, or on dry cat food supplemented with liver. They eat to energy requirements, so on moist cat food they can become protein-deficient, leading to infertility. They can also be fed a complete high-protein dry carnivore diet (32 per cent protein), which is usually soaked for ferrets and fed as a stiff paste. They have little ability to digest fibre.

Ferrets eat 75 g dry diet or 150 g soaked pellets daily.

Food is usually provided in open bowls once daily, and the bowls are cleaned after each meal.

Water

Water is given *ad libitum* from bottles or cups.

Environment

Ferrets are intelligent creatures, and enjoy a varied environment with places to hide and explore. They are agile and like to climb, and branches within the cage are well used. The temperature should be between 15°C and 24°C, and the humidity 45–65 per cent. Good ventilation is required. A 12 hour light cycle is usual.

Breeding

Puberty occurs at 9–12 months, in the spring following birth. Breeding is seasonal. Females are active March to August, and males December to July. This season is light-dependant, oestrus being triggered by increasing day length, and artificial lighting can induce oestrus early and prolong the season.

The vulva of the female enlarges to signal oestrus, and if not mated ovulation does not occur. In this event, persistent oestrus may develop, in which the prolonged high levels of oestrogen can lead to bone marrow depression. To prevent persistent oestrus, it is advisable to spay female ferrets which are not to be bred.

Coitus results in ovulation. If unfertilized, pseudo-pregnancy may develop. Otherwise, gestation lasts 42 days, and 7–8 kits are born. Weaning occurs at 6–7 weeks. If bred early in the year, a female may have two litters in a year.

Growth

Kits are born at 7–10g. Adult weight is reached by 4 months of age, and males are twice the size of females.

PRIMATES

The order Primates includes a wide variety of mammals. Two suborders are recognized:

> Prosimians or subprimates, which includes tree shrews and lemurs;
>
> Anthropoidea or true primates, which includes monkeys, apes, and man.

The most commonly used primates are monkeys.

The Anthropoidea divide into Old World monkeys, the Cercopithecoidea, from Africa and Asia; and New World monkeys, the Ceboidea, from South America. These groups differ in biological needs, and will be dealt with separately. See Figure 6.12 for a simplified classification of Primates.

Primates are intelligent creatures, and they live in groups with well-developed social structures. They communicate in a variety of ways with each other, and in the laboratory with humans. They use sounds, facial expression, and postural changes. Many messages can be easily interpreted. For example, shaking a tree is a sign of aggression, which translates in a captive environment to rattling the cage. Grimacing is a sign of appeasement, which humans often interpret as a smile!

Primates are dextrous and agile, and their environment should be designed with these abilities in mind. They have a great ability to

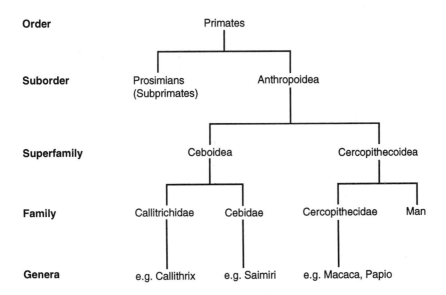

6.12 Simplified classification of Primates.

undo locks and dismantle cages, and will readily remove badly designed food hoppers to escape. Cages should be secure with suitable locks, and they should be made of smooth impervious material such as stainless steel. Usually mesh floors are provided, with trays beneath which are flushed down daily. Power hoses promote the formation of aerosols which may allow transmission of zoonotic diseases, so suitable protective clothing must be worn when operating these.

The potential of primates to carry fatal zoonotic diseases should not be under-estimated. Captive-bred animals are less of a risk, and these should now comprise the majority of animals used. However, primates should not be handled unless sedated. To facilitate the administration of sedatives, most cages have crush-mechanisms. The animal can be held against the front or rear of the cage and an injection given through the bars. For the smaller New World monkeys, leather gloves may be used to catch the animal prior to giving the sedative. Gloves should always be worn to handle primates.

A lack of stimulation rapidly results in boredom for most primates, and this soon leads to stereotypic behaviour. Environmental enrichment is particularly important in monkey houses. Toys, forage trays,

6.13 Gringo receiving titbits.

social contact, mutual grooming, play, sounds, branches to climb and chew, new faces, and new places all add to the quality of the environment and reduce stereotyping (see Figure 6.13).

Primates will eat a variety of foods, including fruit, nuts, insects, and commercially prepared feeds. The vitamin content is particularly important. Primates cannot synthesize vitamin C, and lack of it delays wound healing and predisposes to infections. Classical scurvy will develop if deprived of vitamin C for long periods. It should be supplemented at 1–4 mg/kg bodyweight. Commercial diets with added vitamin C need to be used within 90 days of manufacture or the level of active vitamin declines.

New World monkeys require vitamin D_3 which they normally produce during exposure to ultraviolet light. In the laboratory, it should be added at 1–2.5 iu per gram of diet. A deficiency in it will predispose to pathological lesions of the bones.

Protein should be fed at 3 g/kg bodyweight daily. Diets for New World monkeys usually contain 20–25 per cent, and for Old World monkeys 15–25 per cent.

Primates usually do well with a low dietary fat content, typically 5 per cent.

All dietary needs can be supplied with a commercial pelleted diet, but the addition of fruit and tit-bits can add to the richness of the environment. A small quantity of food mixed with shavings and spread on the floor can keep monkeys occupied for considerable periods of time, and allows them to practise dexterity and discrimination skills (see Figure 6.14).

6.14 Group housed rhesus monkeys at the foraging tray.

NEW WORLD MONKEYS

These originate from tropical rain forests in South America, where they inhabit the upper canopy of the trees. The environment there is hot and humid, and the animals are exposed to sunlight. They are generally arboreal, and climb well. If startled, they flee upwards, not along the ground, and cages for these species should be tall and narrow to reflect this. For larger species, cages may be over 1.5 m in height.

The environment should be maintained at temperatures between 20°C and 28°C, with humidity of 55–65 per cent. Ultraviolet light should be supplied if possible but, if not, vitamin D_3 must be added to the diet.

Two genera are commonly used in research, the common marmoset, *Callithrix jacchus*, and the squirrel monkey, *Saimiri* species.

Marmoset
Behaviour
Wild marmosets live in family groups of 3–8 animals or more. Their social structure is well developed. Males and females pair for life, and both parents care for the young. Individuals show a great deal of interaction, including play and mutual grooming. They are active diurnally.

Housing
Marmosets are usually housed in family groups composed of an adult pair and one to three offspring. They are agile and like to climb, and so benefit from branches or other suitable perches within the cage. They will walk along branches on all four feet and use their tails to balance.

Feeding
Marmosets will eat freshly milled New World monkey diet, but benefit from supplements. Vitamins must be supplied as mentioned above.

Marmosets eat 20g pelleted diet daily.

Food hoppers and water bottles can be cleaned twice weekly. Water is supplied *ad libitum*, using bottles or automatic systems. With bottles, vitamins can be added to the water.

Breeding
Marmosets, as all New World monkeys, have oestrous cycles, whereas Old World monkeys have menstrual cycles. Sexual maturity is reached at 14 months, then the female is polyoestrous, having cycles every 14–18 days. However, no overt signs of oestrus are seen. The male and female form a monogamous pair. If they are

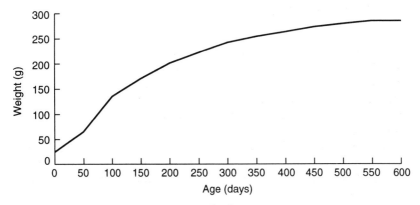

6.15 Typical marmoset growth chart.

kept in large groups with more than one female, only the dominant female will breed.

Mating results in gestation lasting 140 days. Two young are usually born and they stay with their parents until sexual maturity. Females are mated again shortly after parturition and the interbirth interval is 154–178 days. The young are normally weaned at 6 months.

Growth
Males and females exhibit a similar pattern of growth (see Figure 6.15).

Squirrel monkeys
Behaviour
Squirrel monkeys live in large groups of both sexes, but males and females only interact in the breeding season, which in the northern hemisphere is March to May. During this time, aggression between males may be seen, and a dominance hierarchy is apparent. Little mutual grooming is observed between individuals, but they do sleep huddled together.

Housing
Squirrel monkeys can be group housed. A stable hierarchy is soon established, and fighting is only seen if new males are added or in the breeding season. Tall cages are usually used, as mentioned previously, and branches are added.

Feeding
Squirrel monkeys require a high–energy diet, with extra vitamin C. It should be added at 200 mg/kg diet. Neonatal monkeys also have a

particularly high protein requirement. **Squirrel monkeys eat 45–60g pelleted diet daily.**

Water
Water is usually *ad libitum* from bottles or automated systems. **Water requirement is 100–300 ml/kg for adults, and 500 ml/kg for infants.**

Breeding
Squirrel monkeys are usually bred in a harem system with 1 male and up to 12 females, or a multi-male multi-female system. Puberty occurs at 1.5 years in the female, but they are not normally bred until 2.5 years. Males reach puberty at 3.5 years.

Females are seasonally polyoestrous, and cycle every 7–13 days. Mating occurs March to May, and gestation lasts 168–182 days. One infant is usually born.

Little parental care is given to the infant. It will cling to its mother's back, and crawl round to the front to feed.

Growth
They can be successfully hand–reared, and they should gain 20 g each week for the first 8 weeks of life. See Figure 6.16 for the squirrel monkey growth chart.

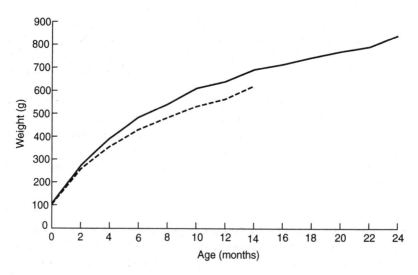

6.16 Typical squirrel monkey growth chart. ——, male; -------, female.

OLD WORLD MONKEYS

Old World monkeys live in a variety of habitats in Africa and Asia, and are partly terrestrial and partly arboreal. Generally, they roam and forage on the ground, and sleep in trees. They have calluses on their ischial tuberosities to facilitate sitting on narrow branches. Their housing usually provides relatively less height than New World monkey caging to reflect the difference in habits.

A wide variety of environmental conditions can be tolerated by these monkeys, but they thrive between 15°C and 24°C, with humidity at 45–65 per cent. Draughtless ventilation providing 12–15 air changes per hour is adequate.

MACAQUES

The macaques most commonly used in research are the rhesus monkey (*Macaca mulatta*), and the cynomolgus or crab-eating monkey (*Macaca fascicularis*). Rhesus monkeys are known for living near human habitation, which they raid for food, and tend to be terrestrial. Cynomolgus monkeys are found more in coastal areas and will eat crustaceans.

Behaviour

Macaques are quadripedal, but will walk upright if carrying infants or food. They are very sociable, and live in large groups, usually with several males. These form a dominance hierarchy, and are responsible for leading the group and imposing discipline. Females have a less marked hierarchy. They are responsible for rearing the young, then maintain a bond with their offspring to keep the group together. Individual relationships within the group depend on kinship and on personalities. Dominant males and females keep a central position in the group, with subordinate males on the periphery to protect them.

Juvenile macaques play together in age groups. Some play is obviously practice for adult life, but some appears to be just for fun. Social contact and mutual grooming are also seen. Playing and normal behaviour are essential preparations for adulthood. The inability to play and express natural behaviour rapidly results in stereotypy.

Housing

Macaques can be kept in outdoor gang cages if well accustomed to a captive environment, but more usually are kept indoors. Quarantine animals may be kept singly or in pairs, in cages with one or more solid sides to reduce aerosol spread, but all other animals benefit from group housing. Compatible animals, such as a disease-free breeding colony, are kept together in large gang cages. Some cages are designed to be joined together, and great flexibility can be achieved using these. Care must be taken though in adding animals to established groups in case fighting occurs.

Feeding

Macaques require approximately 420 kJ/kg daily for maintanance, 525-630 kJ/kg if pregnant or lactating, and 840 kJ/kg if neonatal. commercial Old World monkey diets are used, with supplements of fruit and nuts. Food should be given twice daily to prevent bloat. Foraging mix (see Figure 6.14) can be made up from a selection of:

  - peanuts in the shell
  - flaked maize
  - dog biscuits
  - locust beans
  - sunflower seeds
  - pine kernels
  - boiled eggs in the shell
  - dog chocolate drops

Water

Rhesus monkeys need 80 ml/kg/day. It is usually provided *ad libitum* in bottles or automated system.

Breeding

Puberty occurs at 2–3 years for female rhesus monkeys, and 3–4 years for males. For cynomolgus monkeys, puberty occurs at 3–4 years in both sexes.

Rhesus monkeys have a menstrual cycle lasting 28 days, and breed seasonally from September to January. Mating occurs several times during the receptive period. Gestation lasts 164 days, and 1 infant is usually born. Lactation is prolonged, and weaning occurs after 7–14 months.

Cynomolgus monkeys are non-seasonal breeders, with a menstrual cycle of 31 days. Gestation lasts 167 days, and the single infant is usually weaned at 14–18 months.

Growth

Old World monkeys show a prolonged period of development prior to adulthood. Their life can be divided into phases, eg. for rhesus monkeys:

Fetal phase	164–167 days
Infantile phase	9 months
Juvenile phase	1 year 9 months
Adult phase	11 years plus

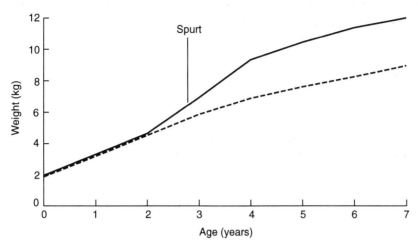

6.17 Typical rhesus monkey growth chart showing adolescent growth spurt. ——, male; -------, female.

This prolonged period of immaturity allows for a long period of learning, but with no sexual competition. Therefore there is little discord in the group while the juveniles are learning from their parents.

Both sexes show an adolescent growth spurt, particularly males (see Figure 6.17).

Baboons

Laboratory baboons are derived from *Papio cynocephalus*, the savannah baboon from Africa. This species encompasses four sub-groups that were previously classified separately.

Baboons are ground-living primates known for their dog-like faces and their sexual skin, which becomes bright pink during oestrus which is the sexually receptive period. They occupy a wide range of habitats, but *P.cynocephalus* is the most arboreal, tending to sleep in trees.

Behaviour

Baboons live in large troops of 40–80, of all ages, in the savannah. Their behaviour depends on the habitat. In open country, social structure is more marked than for forest dwellers.

Males and females show dominance hierarchies. Subordinate animals present their sexual skin to dominants, which may mount the subordinate to reaffirm dominance. Males lead the troop, impose discipline and defend against attack, and females rear young. In a crisis, dominant males stay central within the troop and defend the females, whereas subordinate males sit on the periphery as guards.

Activity tends to be diurnal, and there is much interaction between animals, including mutual grooming and play.

Housing

Baboons are often kept in gang cages in family groups, consisting of 1 male and 1–4 females and offspring, which stay with their parents until adulthood. They need perches, as they sleep in trees.

Feeding and water

Baboons are omnivorous, and will eat fruit, grasses, roots, lizards, and insects. Commercial foods are available. They need 260 kJ/kg if adult, and 1218 kJ/kg if neonatal.

Water is supplied *ad libitum*.

Environment

Environmental enrichment is particularly important, to prevent aggression. Perches, visual barriers, and hiding places should be provided.

Breeding

Females reach puberty at 3.5–4 years, and males at 4–6 years. The menstrual cycle of the female lasts 35 days. Mating occurs at oestrus when the sexual skin is bright pink. Only dominant males are accepted at peak receptivity, but subordinate males may mount the females prior to this. Most sexual activity occurs in the morning, and breeding is non-seasonal. Gestation lasts 170 days. A single infant is born, and is weaned at 5-8 months.

LARGER DOMESTIC SPECIES

The larger species, such as sheep, goats, and pigs, are more familiar to many people than the smaller laboratory species. Air conditioning may not be necessary provided there is adequate draught-free ventilation and the animals are kept dry. Generally, their environment does not need to be as closely controlled as for the smaller species.

Pig

Domestic pigs are believed to be derived from the European wild boar, *Sus scrofa*, crossed with Asiatic pigs such as *S. vittatus*. Many breeds are available nowadays, with different characteristics rendering them suitable for bacon or pork production. Pigs in the laboratory are often ex-breeding sows, and common breeds include the Large White and the Landrace. Miniature pigs are also popular, and are easier to keep in a laboratory due to their smaller size. A commonly used miniature pig is the Yucatan.

Behaviour

Pigs are lively, gregarious animals which are usually docile, although adult boars should be approached with caution. They form stable groups, but there is a tendency to fight if unfamiliar animals are placed together. Tail-biting and navel sucking are common results of mixing, particularly if done in confined spaces. To reduce this, pigs should be mixed on neutral territory at feeding time or at dusk. Tranquillizers may be used in extreme cases.

Pigs are intelligent, and will rapidly become bored with a sterile, uninteresting environment. Stereotypic behaviour patterns such as bar-biting will then be seen. Group housing, toys such as chains plastic bottles, and boxes, and bedding to root in all add to environmental enrichment and reduce stereotypy (see Figure 6.1).

Housing

Pigs are usually kept indoors in draught-free buildings. Groups of up to 10 can be kept together. To prevent fighting, new animals should not be introduced to the group.

Pigs are usually kept on solid or slatted concrete floors for hygiene. Slats should be designed to a particularly high standard to prevent damage to the feet. Bedding such as sawdust, shavings or straw, is usually provided on solid floors, where it will provide warmth and comfort, and somewhere to root. Straw will be chewed and played with by the pig, helping digestion and reducing boredom. If no bedding is provided more leg injuries will occur, and the ambient temperature must be kept higher.

Although bedding may appear to interfere with cleaning out procedures, a pig with sufficient space will dung only in the coolest and wettest area, so just this area needs to be cleaned daily. Pig accommodation may be designed with an indoor kennel with a solid floor and bedding, and an outdoor solid or slatted area which can be hosed or scraped daily.

Different accommodation is required for farrowing pigs. Overlying of piglets often occurs, so sows may be kept in farrowing crates between parturition and weaning. These stop the sow from lying down quickly, so piglets trapped underneath have a chance to escape. Also, newborn piglets require a different micro-environment from their mothers. They need a temperature close to 30 or 32°C, so creep areas adjacent to the sows udder with a nest box, heat lamp, and creep feeder should be provided.

All pig accommodation needs to be cleaned out daily, and disinfected thoroughly each month.

Pigs can be kept outdoors in groups with several compatible

females and males. Huts with plenty of straw should be provided for these to protect them from cold in the winter and sunburn in the summer.

Feeding
Pigs are omnivorous and the nutritional requirements vary with the age and physiological status of the pig.

Creep feed is fed to suckling and newly weaned pigs. The protein level is usually over 20 per cent. From 2 weeks of age, piglets require more than milk, so creep feed is generally introduced at an early age. To prevent iron deficiency, an injection of iron dextran is usually given at birth. Creep feed is given *ad libitum* until the piglets are 25–30 kg in weight, then restricted.

For production, a diet with 16–18 per cent protein is fed. Production rations are given in the last trimester of pregnancy and during lactation. They are introduced gradually, and should be given *ad libitum* for the period of maximum production.

For maintenance, a diet with 12 per cent protein and plenty of fibre is fed. Pigs can easily become obese on a maintenance diet, and should be fed restricted amounts, with extra fibre for bulk. Any changes in diet must be done gradually to allow time for the gut flora to adapt.

Fat is usually kept low, at 3–3.5 per cent in the diet.

Standard diets are available from merchants, to be fed alone or with extra roughage. They are fed dry if automatic watering is available, or mixed to a gruel if not.

Breeding animals are fed twice daily, and other animals once daily.

Miniature pigs have similar feeding requirements to large pigs, but with quantities reduced in proportion to their bodyweight.

Water
Water is particularly important for pigs, and must be available at all times. Automatic watering systems may be used, or if not the food should be given wet then troughs refilled with water when feeding has finished. Lack of water leads to salt poisoning, which is rapidly fatal. **Water should be supplied at twice or three times the amount of dry food given.**

Environment
Pigs need to be protected from extremes of temperature. If the building is well insulated and draught-free, and bedding is provided, then temperatures above 10°C are well tolerated. The optimum temperature is between 16°C and 18°C for adults, and 30° and 32°C for neonates. Humidity should be kept between 60 per cent and 70 per cent to reduce the incidence of respiratory diseases, but the main

factor in control of these is ventilation. Adequate ventilation must be supplied to reduce the levels of ammonia and pathogens in the air, but air speed should not exceed 0.2–0.3 m/s for adults and 0.1 m/s for piglets or excessive cooling occurs.

Breeding

Pigs are prolific breeders and will breed all year round.

For Large White pigs, puberty occurs at 6 months. The oestrous cycle lasts 21 days, and mating results in a pregnancy of 114–115 days. An average of 10 piglets is born, weighing approximately 1.3 kg each. Weaning occurs at 3–8 weeks (early weaning is favoured), and the sow returns to oestrus 7 days after weaning.

For miniature pigs, puberty occurs a little earlier, at 4 months. The oestrous cycle is 19.5 days, and pregnancy lasts 113–114 days. Piglets are born weighing approximately 0.5 kg, and are weaned at 7 weeks.

Post weaning sows and post-pubertal gilts should be kept in groups within the sight and smell of the boar. His presence and the 'dormitory effect' of other females leads to oestrus. Sows are either taken to the boar for mating or inseminated artificially. Once pregnancy has been confirmed, sows are usually penned separately until farrowing is due. Then they are bathed and checked over prior to being put in the farrowing house. Accommodation in these is usually in single pens with creep areas for the piglets and bedding. Farrowing crates may be used.

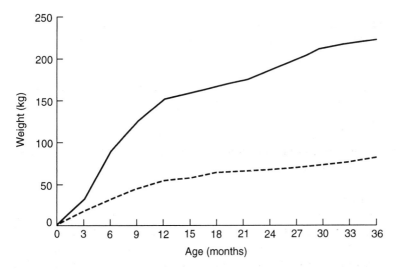

6.18 Typical Large White pig and Yucatan pig growth chart.
——, Large White; -------, Yucatan.

Growth

Piglets are born quite immature, but with eyes and ears open. Colostrum from the sow is essential as no antibodies cross the placenta prior to parturition. They must therefore be ingested in the colostrum within the first 24 hours, after which time intestinal closure will prevent absorption of antibody. Initially piglets are suckled hourly, then the frequency declines. As piglets are born with teeth, these are often clipped out to prevent damage to the sows udder.

Piglets should weigh 5 kg when weaned at 3–4 weeks, and the daily weight gain should be 150 g. Between 40 kg and 100 kg, the daily gain should be 700–800 g. (See Figure 6.18.).

Sheep and goat

Sheep (*Ovis aries*) and goats (*Capra hircus*) are generally kept for meat, hide, and milk, but also have a wide variety of uses in experiments, frequently for research into meat and milk production.

Many breeds of sheep and goats are available, with differing temperaments, sizes, and fleece/coat types, with and without horns.

Behaviour

Sheep are grazing animals. Naturally, they rely on numbers for safety and like to remain in large groups. They are easily startled and tend to huddle together if they suspect danger.

Goats are browsing animals and tend to be more independent. They use their agility and rapid reactions to avoid danger. They are very curious, and will stand on their hind feet to explore. This way, large goats can reach a height of 2 metres, so this should be considered when suspending items such as cables near or above goat pens.

Both sheep and goats can thrive on poor land, but goats particularly can live on very steep hills, where sheep cannot survive.

Housing

Group housing is best for sheep and goats, but if individual penning is required, they should be able to see each other. Rams tend to be housed separately from ewes except during the breeding season, and ewes are often grouped according to age and condition to facilitate differential feeding.

Goats can be kept in the same way. Polled animals of either species should be kept separate from horned animals.

Housing should be constructed with smooth, impervious walls and floors. Bedding may be provided in the form of straw or sawdust, and needs to be deep enough to keep the fleece dry. This also helps provide insulation. Alternatively, slatted floors may be used. Pens should be scraped and washed down every 2–3 days, and allowed to dry before the animals return.

Sheep and goats are frequently kept outdoors. Sheep will remain outdoors in all weathers but goats dislike rain and need a shelter in the field. This can be an open fronted building provided it backs into the prevailing wind. Good fences are essential to prevent sheep and goats from escaping.

Ewes and nanny goats are brought indoors for lambing/kidding into partly or fully covered buildings. Ventilation must be good to reduce ammonia levels, and well-designed housing allows this without the need for mechanical fans. Sheep are kept in groups of up to 30 indoors; in larger groups there is danger of crushing. Ewes are moved to individual pens for lambing.

Once the maternal bond is established (4–7 days) and lambs have received colostrum, ewes and lambs are returned to the field. Ewes with lambs at foot are kept separate other ewes.

Food and water

Sheep and goats are ruminants: the forestomach has developed into three large chambers containing microbes which digest plant products and manufacture many vitamins. The rumen is fully functional by 3–4 months of age.

During grazing, food passes into the rumen. Then the animal rests and food is regurgitated for chewing, before being returned to the rumen for digestion by the microbes. Most nutrients are absorbed directly in the rumen. Sheep and goats thrive on a diet of grass products with minimal use of concentrates. They require cellulose from grass, hay silage or straw, and a supply of cereal products. Mineral and vitamin licks may be supplied, but copper levels must be carefully controlled as an excess is especially toxic for sheep and goats.

Good quality hay is sufficient for maintenance. Goats have a higher metabolic rate than sheep, so require greater quantities, but can utilize a wider range of feedstuffs.

Concentrates are not generally fed to sheep and goats, except for the last trimester of pregnancy (the 'steaming up' period), and during lactation. Concentrates designed for sheep specifically should be used, not those for cattle. Approximately 100 g is fed daily from day 100 of pregnancy, and increased to 400 g or more throughout lactation. Food and water can be supplied in troughs or hoppers fitted to the outside of the pen door, hay can also be given in overhead racks. Sufficient trough space must be provided per head to prevent competition and bullying, and troughs must be high enough to avoid being contaminated with faeces or bedding.

Each animal must be provided with 4–6 litres of water daily:

lactating goats can drink more than 6 litres in a day. Water troughs must be cleaned frequently, as goats can be fussy if water is dirty.

Environment

The temperature for sheep and goats indoors should be between 10°C and 24°C, but this depends on the fleece length. Ventilation is important—extractor fans should remove 3 m³ air per kg bodyweight per hour to keep ammonia levels down.

Often sheep and goats are kept outdoors in a farm environment, in which case, provided accommodation is dry and draught-free, it is acceptable.

Breeding

Sheep and goats are both seasonally polyoestrous.

Sheep reach puberty at 7–12 months, then have oestrous cycles lasting 16.5 days between September and March. Ewes and rams are put together in these months, 40–50 ewes per ram or 25–30 per ram lamb. Gestation lasts 140–150 days. Lambs are born in the spring—usually twins. Weaning time is variable depending on the flock requirements, but it is usually done at 16–20 weeks, when the rumen is functional.

Goats are polyoestrous between September and January. Males are run with females during the season, or females can be taken to the male individually at oestrus. The cycle lasts 21 days. Pregnancy lasts 144–157 days in goats and detection can be difficult. One to three kids may be born, mainly in March.

Oestrus can be synchronized, in sheep particularly, for example by using progesterone sponges or vasectomized rams, and the breeding season can be advanced by the use of melatonin.

Feeding is increased prior to the breeding season, in the so called 'flushing' period, to improve body condition and increase conception. Good feeding should continue through the first third of pregnancy to reduce embryonic mortality.

Parturition may take place within the group, or females about to give birth can be penned individually.

Growth

Lambs and kids should double their birthweight by 3–3.5 weeks, reach 70 per cent of adult weight at one year, and be fully grown at 2–3 years.

7 Disease prevention and health monitoring in animals

If it is recognized that an animal is unwell but that the changes are unrelated to the procedure, it will be necessary to consult the named veterinary surgeon in order to investigate the causes of the disease and obtain advice on treatment.

It is important to prevent disease occurring in laboratory animals, not only for the benefit of the animal, but because the effects of disease can confound experimental results. It is therefore customary to monitor the health of animals used in laboratories. The researcher should be aware of the pathogen status of the animals used, not just initially, but throughout the course of study. The animals may not show clinical signs when they arrive but may harbour pathogens which are capable of severely compromising the health of the animal when it is subjected to experimental stress. Alternatively, the infection may not cause clinical disease but may induce microscopic or biochemical changes which have profound effects on research data.

SOURCES OF CONTAMINATION

Rodents, and many other species, are generally supplied with a health profile from the breeder. If the animals are to be kept short term only it may not be necessary to monitor, but if they are kept long term, continual regular screening will be necessary to ensure they have not picked up extra infections. These infections could come from other animals or from the environment, so it is necessary to consider the husbandry under which the animal is kept. Animals may be housed in conventional units or barrier units. The latter may be for *axenic* animals (full barrier environment with Caesarean rederivation resulting in germ-free animals), *gnotobiotic* animals (with a defined flora present) or *specific pathogen-free* animals (part barrier to keep the animal free of a known range of microorganisms). If the animal is kept under conditions to prevent entry of disease it is important to maintain these barriers and not to bring microorganisms into the unit.

To ensure research animals remain free from disease it is necessary to consider the potential sources of infection:

1. *Other animals.* Consider where other animals in the unit are coming from. Examine the supplier's health profile and ask what tests were carried out, who did them and how the interpretation was arrived at. It is surprising how often two batches of serum from the same animals, sent to two different laboratories will come back with two different answers on the disease profile.

2. *Water.* Find out if the water supply for the animals is purified, and if so, how this is carried out, and what methods are used for checking it has been done effectively.

3. *Food.* Find out how the food that is going to be given to your animals is being stored, and check that there is adequate protection from contamination by wild rodents or birds which may introduce disease (e.g. *Salmonella*). Ensure that the method of storage prevents depletion of vitamin levels or the development of moulds or toxins.

4. *Bedding.* Look at the quality of the bedding and ask how the microbiological quality is assured.

5. *Air.* Ensure that the ventilation rates comply with the Home Office Code of Practice and are adequate to remove any build up of ammonia from within the room which will predispose to respiratory disease. Find out how often the air filters are changed and ensure this is done frequently enough.

6. *Other disease vectors.* Check that all your equipment is clean and will not spread disease to your animal. Examples of potential sources of contamination include:

   – anaesthetic face masks
   – clippers
   – gloves/dirty hands
   – white coats
   – hypodermic needles (use once only)
   – gavage cannulae
   – coughs and sneezes (wear a mask)

Some animals may have predisposing factors such as a genetic variant which contribute to the development of disease. For example, BALB/c mice are more susceptible to lethal ectromelia infection than the C57BL/6 strain. Genetic monitoring of inbred colonies is vital to ensure that this variable predisposition remains a constant factor within the experimental model.

Nutritional status of the animal will also be a factor in susceptibility to disease and must be considered carefully.

However, the single most important predisposing factor in the laboratory animal is STRESS which leads to immunosuppression and the subsequent development of disease. It is therefore very important to do everything possible to reduce the stress caused to your animals.

HEALTH MONITORING

The health of your animals can be monitored passively or actively. Passive monitoring is studying information which is readily available from the population, without actually killing animals simply for the purpose of health screening. All animals which die unexpectedly, all animals which produce unusual experimental results, and a proportion of all animals used terminally should be examined routinely. The scope of this examination will depend on the facilities readily available, but the minimum will be a gross post-mortem. It may be possible, and is certainly desirable, to add to this routine bacteriology, parasitology, serology, and histology. A routine faecal screen from the live animal can also be useful in detecting various subclinical infections.

In active health monitoring, animals are taken from the population at regular intervals. Blood is taken from them for serological testing for a number of viral infections and they undergo full post-mortem examination to detect any other infections. Sentinel animals can also be used to monitor the population. These are animals of the same species whose bacterial, parasitic, and virological burden is already defined. They are housed in cages on the lower shelves of the racks of the experimental population, and exposed to their dirty bedding. They will then take on the same pathogens. They can be sampled to determine their pathogen status which will then reflect the status of the original experimental population.

TECHNIQUE FOR ROUTINE POST-MORTEM EXAMINATION

If a post-mortem examination is to be of any value then it must be conducted thoroughly. The practice of simply slitting open the abdomen to have a look at some internal organs without removing them is usually a waste of time. Although it may not be feasible to dissect the carcase minutely, it is desirable to examine the whole cadaver and remove and inspect each organ. It may then be necessary to submit portions of diseased tissues to a laboratory for further investigation. The basic procedure is the same for all species with minor variations depending on the size of the animal involved.

Procedure

1. Weigh the animal and examine the external appearance of the carcase. Make a note of the body condition, and examine all external orifices for any abnormalities or discharges. Examine the feet, tail, and skin for any lesions and note any evidence of vomiting, diarrhoea or dehydration.

2. Lay the animal out on its back, pin the feet if necessary, or incise the axillae and over the femoral heads so the limbs can be forced downwards to support the carcase. Swab the ventral surface liberally with disinfectant to reduce bacterial contamination and prevent pollution of the area with dander and fur.

3. Make a midline skin incision from mandible to pubis and reflect the skin away from the midline. Remove the muscles over the thorax and open the chest cavity by holding the xiphisternum and cutting through the costochondral junctions on each side. Reflect the sternum and ribs anteriorly to expose the underlying viscera. Observe any abnormalities such as fluid in the pleural cavity or pericardium. Open the trachea and note any inflammation or fluid inside.

4. Open the abdominal wall with a straight midline incision and reflect the wall sideways. Again, note any obvious changes such as liver abscessation or fluid in the peritoneal cavity.

5. Grasp the trachea and oesophagus, cut across and lift upwards to remove the heart and lungs, easing away any dorsal connections. Cut across the oesophagus as it passes through the diaphragm and remove the thoracic viscera for a more thorough examination on a separate tray. Look for areas of discolouration, altered texture, adhesions or other abnormalities. Open the length of the trachea, oesophagus, and bronchi. Cut through the lung lobes and squeeze gently to see if there is any fluid present in the lung parenchyma. Open the heart by incising from the apex of the left ventricle up into the left atrium and into the aorta, and from the apex of the right ventricle up into the right atrium and into the pulmonary artery. This method ensures that the heart valves and endocardium can be examined undamaged by the dissection.

6. Examine the intestinal tract carefully noting such facts as whether the stomach is full or empty, whether the gut is full of gas, fluid or solid matter or any areas of inflammation. Remove the intestinal tract and examine the mesenteric

lymph nodes and the inside surface of the stomach and intestines.

7. Cut around the dorsal surface of the liver to remove it and cut through the lobes. Examine the cut surface for variations in the normal architecture.

8. Examine the urogenital tract. Incise the kidneys longitudinally and inspect the internal structure. Peel off the capsule.

9. The eviscerated carcase should then be examined for any other abnormalities such as enlarged joints or lymph nodes. If the history indicates it, the head can be removed to take out the brain for detailed inspection.

10. Ensure the carcase and viscera are correctly disposed of and the area thoroughly cleaned and disinfected.

WHICH DISEASES TO MONITOR

It is obviously not practical to screen for every possible disease in your experimental animal, so a decision must be made about which ones to monitor. The threat of disease should be perceived as serious by all those involved in order for screening to be carried out effectively. There are three groups of diseases to consider:

1. The first priority is to screen for any *zoonotic diseases* which may occur in the species being used, bearing in mind their original source. This will enable any necessary precautions to be taken to prevent spread of disease to humans.

2. The second group of diseases to look for will be those that might be present as *subclinical infections but which will cause overt clinical disease* when the animal is subjected to an experimental stress.

3. The third group to consider are those diseases whose presence, whether clinically evident or only subclinical, will *interfere with the experiment*. For example, this may be the presence of a viral infection which has no effect on the animal but elevates certain enzyme levels, the measurement of which may form part of the experimental study.

A discussion with your named veterinary surgeon will aid in designing a sensible screening programme which will be cost-effective. Before embarking on a screening programme it is important to be prepared for the possibility of positive results, and to consider what action will need to be taken if the result shows the presence of certain infections. The epidemiology of the disease must be understood in order to know how frequently to test, what samples to take, and how

to deal with a positive result if it should be found. The programme
will need to be re-evaluated once the status of the colony is known, as
subsequent tests may need to be done less frequently, or only certain
groups of the population tested, after the initial screen. Screening can
be a very costly and laborious task but this should not be used as an
excuse to ignore the possibility that infections exist which may be
of concern to personnel, detrimental to the animals' welfare, and to
the validity of the science being carried out.

THE LABORATORY ANIMALS BREEDERS ASSOCIATION ACCREDITATION SCHEME (LABAAS)

This scheme covers the different methods of breeding species which
are normally purpose-bred for use in the laboratory (i.e. mouse,
rat, guinea-pig, gerbil, hamster, rabbit, dog, cat, and ferret), and
the supply of those animals which are not usually purpose-bred.
There are four classifications of accreditation available:

1. Full barrier breeder
2. Part barrier breeder
3. Non-barrier breeder
4. Supplier

Accreditation is awarded to LABA members who meet the required
standards of facilities, husbandry, and management and who operate
a health monitoring programme which meets the requirements of the
scheme. See Table 7 for the requirements for each species.

Table 7

MICROBIOLOGICAL SCREENING NECESSARY TO CONFORM WITH LABAAS REQUIREMENTS

MOUSE

Viral infections

List of viral infections to be serologically monitored in mouse breeding units

No.   Antigens

M1   Minute virus of mice (MVM)
M2   Mouse hepatitis virus (MHV)
M3   Pneumonia virus of mice (PVM)
M4   Reovirus type 3 (Reo3)
M5   Sendai virus
M6   Theiler's encephalomyelitis virus (TMEV/GDVII)

M7   Hanta viruses
M8   Lymphocytic choriomeningitis virus (LCM)
M9   Ectromelia virus
M10   Lactate dehydrogenase virus (LDV)*

M11   Mouse adenovirus (MAd)**
M12   Mouse pneumonitis virus (K)**
M13   Mouse polyoma virus**
M14   Mouse thymic virus (MTV)**
M15   Mouse rotavirus (EDIM)**
M16   Mouse cytomegalovirus (MCMV)**

Bacterial, mycoplasmal, and fungal organisms

List of bacterial, mycoplasmal and fungal organisms to be monitored in mouse breeding colonies

Bacillus piliformis
Bordetella bronchiseptica
Citrobacter freundii (4280)
Corynebacterium kutscheri
Pasteurella multocida
Pasteurella pneumotropica
Salmonella spp.
Streptobacillus moniliformis
Streptococci-β-haemolytic (except D group) (Designation of Lancefield Group, if possible)
Streptococcus pneumoniae
Mycoplasma pulmonis
Dermatophytes

Parasite infection/infestation
Intestinal protozoa
Helminths
Arthropods

* Raised enzymatic activity
** Evidence exists for rare infections of these viruses in mouse colonies. They should be tested in rederived or restocked colonies.
For test schedule mouse breeding colonies, see p. 108.

RAT

Viral infections
List of viral infections to be sero-logically monitored in rat breed-ing units

No.   Antigens

R1   Hantaan virus
R2   Kilham rat virus (KRV)
R3   Pneumonia virus of mice (PVM)
R4   Reovirus type 3 (Reo3)
R5   Sendai virus
R6   Sialodacryoadenitis virus
      (SDA)/Rat Corona virus
      (RCV)
R7   Theiler's encephalomyelitis
      virus
      (TMEV/GDVII)
R8   Toolan's (H-1)
R9   Lymphocytic choriomeningitis
      virus (LCM)

Bacterial, mycoplasmal, and fungal organisms
List of bacterial, mycoplasmal and fungal organisms to be monitored in rat breeding colonies

Bacillus piliformis
Bordetella bronchiseptica
Corynebacterium kutscheri
*Leptospira (icterohaemorrhagiae
   and ballum)*
Pasteurella multocida
Pasteurella pneumotropica
Salmonella spp.
Streptobacillus moniliformis
Streptococci-β-haemolytic
   (except D Group) (Designation of
   Lancefield Group, if possible)
Streptococcus pneumoniae
Mycoplasma pulmonis
Mycoplasma arthritidis
Dermatophytes

Parasite infection/infestation
Intestinal protozoa
Helminths
Arthropods

For test schedule in rat breeding colonies, see p. 108.

GUINEA-PIG
Viral infections

List of viral infections to be sero-logically monitored in guinea-pig breeding units

No.	Antigens
GP1	Sendai virus
GP2	Lymphocytic choriomeningitis virus (LCM)

Bacterial and fungal organisms

List of bacterial and fungal organisms to be monitored in guinea-pig breeding colonies

Bacillus piliformis
Bordetella bronchiseptica
Pasteurella multocida
Pasteurella pneumotropica
Salmonella spp.
Streptobacillus moniliformis
Streptococci-β-haemolytic
   (Except D Group) (Designation of
   Lancefield Group, if possible)
Streptococcus pneumoniae
Yersinia pseudotuberculosis
Dermatophytes

Parasite infection/infestation

Intestinal protozoa
Helminths
Arthropods

For test schedule in guinea-pig breeding colonies, see p. 108.

Test schedule in mouse, rat, and guinea-pig breeding colonies

	Mouse	Rat	Guinea-pig
Sampling frequency			
Viruses	Every 3 months Antigen nos. M1–M6	Every 3 months Antigen nos. R1–R8	Every 3 months Antigen no. GP1
	Once a year Antigen nos M7–M10	Once a year Antigen no. R9	Once a year Antigen no. GP2
	After restocking or rederivation Antigen nos. M11–M16		
Bacteria, myco- plasma, fungi, and parasites	Every 3 months	Every 3 months Once a year— *Leptospira* spp.	Every 3 months
Sample size			
Serology			
Age	Not less than 10 weeks	Not less than 10 weeks	Not less than 10 weeks
No. of animals	8*	8*	8*
Bacteriology, mycoplasmology, mycology, and parasitology			
Age	4–5 weeks	4–5 weeks	4–5 weeks
No. of animals	4	4	4

* Four of the animals submitted for serology will also be used for bacteriology, mycoplasmology, mycology, and parasitology.

GERBIL AND HAMSTER

Viral infections

List of viral infections to be sero-logically monitored in hamster and gerbil breeding units

No.	Antigens
HG1	Sendai virus
HG2	Lymphocytic choriomeningitis virus (LCM)

Bacterial and fungal organisms

List of bacterial and fungal organ-isms to be monitored in hamster and gerbil breeding colonies

Bacillus piliformis
Bordetella bronchiseptica
Pasteurella multocida
Pasteurella pneumotropica
Salmonella spp.
Streptobacillus moniliformis
Streptococci-β-haemolytic
  (except D group) (Designation of
  Lancefield Group, if possible)
Streptococcus pneumoniae
Dermatophytes

Parasite infection/infestation

Intestinal protozoa
Helminths
Arthropods

Test schedule in gerbil and hamster breeding colonies

Sampling frequency

Viruses	Every 3 months—Antigen no. HG1 (hamster and gerbil)
	Every 3 months—Antigen no. HG2 (hamster) Once a year—Antigen no. HG2 (gerbil)
Bacteria, fungi and parasites	Every 3 months

Sample size

Serology	
Age	Not less than 10 weeks
No. of animals	8*
Bacteriology, mycology and parasitology	
Age	4–5 weeks
No. of animals	4

* Four of the animals submitted for serology will also be used for bacteriology, mycology, and parasitology.

RABBIT

Viral infection

The viral infection to be serologically monitored in rabbit breeding units

No.   Antigens

RB1   Rabbit pox virus (myxomatosis)

Bacterial and fungal organisms

List of bacterial and fungal organisms to be monitored in rabbit breeding colonies

> *Bacillus piliformis*
> *Bordetella bronchiseptica*
> *Pasteurella multocida*
> *Pasteurella pneumotropica*
> *Salmonella* spp.
> Streptococci-β-haemolytic
>   (except D group) (Designation of
>   Lancefield Group, if possible)
> *Yersinia pseudotuberculosis*

Parasite infection/infestation

> *Encephalitozoon cuniculi*
> Intestinal protozoa
> Helminths
> Arthropods

Test schedule in rabbit breeding colonies

Sampling frequency	
Virus	Every 3 months—Antigen no. RB1
Bacteria, fungi, and parasites	Every 3 months

Sample size	All animals will be used for all investigations

Serology	
Age	7–8 weeks
No. of animals	4

Bacteriology, mycology, and parasitology	
Age	Greater than 26 weeks
No. of animals	4

DOG

Viral infections

List of viral infections to be serologically monitored in dog breeding units

No.	Antigens
D1	Canine distemper virus*
D2	Canine hepatitis virus*
D3	Canine parvovirus*

Bacterial and fungal organisms

List of bacterial and fungal organisms to be monitored in dog breeding colonies

Bordetella bronchiseptica
Campylobacter jejuni
Salmonella spp.
*Leptospira canicola**
*Leptospira icterohaemorrhagiae**
Dermatophytes

Parasite infection/infestation

Helminths
Arthropods

* Screening for this organism may be omitted when there is a declared vaccination policy under veterinary supervision in accordance with the manufacturers' recommendations. The vaccination policy must be made known to customers on request.

Test schedule in dog breeding colonies

Sampling frequency	
Viruses, bacteria, fungi, and parasites	Every 3 months
Sample size	8 animals from each breeding unit
Samples to be taken	Plucked hair, faeces Ear swab, pharyngeal swab Blood (when there is no vaccination policy—see above)
Age	Weaners Adults
No. of animals	4 4

CAT

Viral infections

List of viral infections to be sero-
logically monitored in cat breeding
units

No. Antigens

C1 Feline panleucopaenia virus*
C2 Feline calicivirus*
C3 Feline rhinotracheitis virus*
C4 Feline leukaemia virus*

* Screening for this organism may
be omitted when there is a declared
vaccination policy under veterinary
supervision in accordance with the
manufacturers' recommendations.
The vaccination policy must be
made known to customers on request.

Bacterial and fungal organisms

List of bacterial and fungal organisms to be monitored in cat breeding colonies

Bordetella bronchiseptica
Salmonella spp.
Dermatophytes

Parasite infection/infestation

Intestinal protozoa
Toxoplasma gondii
Helminths
Arthropods

Test schedule in cat breeding colonies

Sampling frequency

Viruses, bacteria, fungi, and parasites	Every 3 months

Sample size	8 animals from each breeding unit

Samples to be taken	Plucked hair, faeces Ear swab, pharyngeal swab Blood (when there is no vaccination policy—see above)
Age	10–14 weeks (young adults) Greater than 6 months (breeders)
No. of animals	4 4

FERRET

Viral infection

The viral infection to be serologically monitored in ferret breeding units

No.   Antigens

F1   Canine distemper virus*

* Screening for this organism may be omitted when there is a declared vaccination policy under veterinary supervision. The vaccination policy must be made known to customers on request.

Bacterial and fungal organisms

List of bacterial organisms to be monitored in ferret breeding colonies

Pasteurella multocida
Pasteurella pneumotropica
Salmonella spp.
Dermatophytes

Parasite infection/infestation

Helminths
Arthropods

Test schedule in ferret breeding colonies

Sampling frequency	
Viruses, bacteria, fungi, and parasites	Every 3 months
Sample size	8 animals from each breeding unit
Samples to be taken	Plucked hair, faeces Pharyngeal swab Blood (when there is no vaccination policy—see above)

	Weaners	Adults
Age		
No. of animals	4 (if available)	4*

* 8 adults if weaners unavailable

8 Biological data

....

INTRODUCTION

This chapter gives details of biological, breeding, haematological, and biochemical data for the common laboratory species. It must be remembered that the ranges given are taken from studies of populations in which there is considered to be a normal distribution, and some 5 per cent of **normal** animals will fall outside the reference ranges, some at the bottom and some at the top. For measurements such as breathing and heart rate, the value obtained will depend to a great extent on the psychophysiological state of the animal at the time. If an animal is excited and stressed by capture and restraint, it will have artificially high heart and respiratory rates. Animals which are asleep will have depressed vital signs. Stress also affects some biochemical and haematological measurements. Blood glucose and cortisol levels rise during stress, and the white blood cell counts will alter if the stress is severe enough, usually leading to a neutrophilia, with lymphopenia and eosinopenia. This blood picture is known as a **stress leucogram**.

For biochemical determinations, the exact method used by the laboratory affects the reference ranges, so ideally results should be compared with other results from the same laboratory. Other reference ranges will usually be of the same order of magnitude, and will give an indication if there is gross variation from normality.

It is important to be consistent when taking measurements from an animal, and to use the same technique each time. If laboratory determinations are to be used, the figures will only be significant if compared against the appropriate reference range. An abnormal or unexpected reading should be checked to see if it is internally consistent, that is if it is consistent with other readings from the same animal. For example, if an animal's blood sample shows a total protein level of 10 g/100 ml, an albumin level of 8 g/100 ml, and a globulin level of 6g/100 ml, then this is not possible: one or more of the readings must be incorrect, as the total protein should equal the sum of the albumin and globulin levels.

The licensee should beware of seemingly anomalous results.

Readings should not be interpreted in isolation, the state of the whole animal should be taken into account.

MOUSE

Biological data

Adult weight (g)	Male 20–40
	Female 18–40
Diploid number	40
Food intake	15 g/100 g bw
Water intake	15 ml/100 g bw
Lifespan (years)	1.5–3
Rectal temperature (°C)	38–39
Heart rate/min	310–840
Blood pressure systole (mmHg)	133–160
Blood pressure diastole (mmHg)	90–110
Blood volume (ml/kg)	60–75
Respiratory rate/min	60–220
Tidal volume (ml)	0.18

Breeding data

Puberty (days)	28–49 (average 42)
Age to breed male (days)	70
Age to breed female (days)	60–84
Gestation (days)	19–21
Litter size	4–12
Birth weight (g)	1–1.5
Weaning age (days)	18–21
Oestrous cycle (days)	4–5
Post-partum oestrus	Fertile

Haematological data

RBC ($\times 10^6$/mm^3)	7–12.5
PCV (%)	39–49
Hb (g/100 ml)	10.2–16.6
WBC ($\times 10^3$/mm^3)	6–15
Neutrophils (%)	10–40
Lymphocytes (%)	55–95
Eosinophils (%)	0–4
Monocytes (%)	0.1–3.5
Basophils (%)	0–0.3
Platelets ($\times 10^3$/mm^3)	160–410

Biochemical data

Serum Protein (g/100 ml)	3.5–7.2
Albumin (g/100 ml)	2.5–4.8
Globulin (g/100 ml)	0.6
Glucose (mg/100 ml)	62–175
Blood urea nitrogen (mg/100 ml)	12–28
Creatinine (mg/100 ml)	0.3–1
Total bilirubin (mg/100 ml)	0.1–0.9
Cholesterol (mg/100 ml)	26–82

RAT
Biological data

Adult weight (g)	Male 450–520	
	Female 250–300	
Diploid number	42	
Food intake	10 g/100 g bw	
Water intake	10 ml/100 g bw	
Lifespan (years)	3–4	
Rectal temperature (°C)	36–40	
Heart rate/min	250–450	
Blood pressure systole (mmHg)	84–134	
Blood pressure diastole (mmHg)	60	
Blood volume (ml/kg)	54–70	
Respiratory rate/min	70–115	
Tidal volume (ml)	0.6–2	

Breeding data

Puberty (days)	60–63
Age to breed (days)	65–110
Gestation (days)	20–23
Litter size	6–12
Birth weight (g)	5–6
Weaning age (days)	21
Oestrous cycle (days)	4–5
Post-partum oestrus	Fertile

Haematological data

RBC ($\times 10^6$/mm^3)	7–10
PCV (%)	36–48
Hb (g/100 ml)	11–18
WBC ($\times 10^3$/mm^3)	6–17
Neutrophils (%)	9–34
Lymphocytes (%)	65–85
Eosinophils (%)	0–6
Monocytes (%)	0–5
Basophils (%)	0–1.5
Platelets ($\times 10^3$/mm^3)	500–1300

Biochemical data

Serum Protein (mg/100 ml)	5.6–7.6
Albumin (g/100 ml)	3.8–4.8
Globulin (g/100 ml)	1.8–3
Glucose (mg/100 ml)	50–135
Blood urea nitrogen (mg/100 ml)	15–21
Creatinine (mg/100 ml)	0.2–0.8
Total bilirubin (mg/100 ml)	0.2–0.55
Cholesterol (mg/100 ml)	40–130

GUINEA-PIG
Biological data

Adult weight (g)	Male 850–1200
	Female 700–900
Diploid number	64
Food intake	6 g/100 g bw
Water intake	10 ml/100 g bw
Lifespan (years)	4–8
Rectal temperature (°C)	37.2–40
Heart rate/min	230–380
Blood pressure systole (mmHg)	80–94
Blood pressure diastole (mmHg)	55–58
Blood volume (ml/kg)	69–75
Respiratory rate/min	42–104
Tidal volume (ml)	2.3–5.3

Breeding data

Puberty	Male 60 days
	Female 30 days
Age to breed male	3–4 months, 600–700 g
Age to breed female	2–3 months, 300–450 g
Gestation (days)	59–72
Litter size	2–5
Birth weight(g)	70–100
Weaning age	3–4 weeks
Oestrous cycle	15–17 days
Post-partum oestrus	Fertile

Haematological data

RBC (× 10^6/mm^3)	4.5–7
PCV (%)	37–48
Hb (g/100 ml)	11–15
WBC (× 10^3/mm^3)	7–18
Neutrophils (%)	28–44
Lymphocytes (%)	39–72
Eosinophils (%)	1–5
Monocytes (%)	3–12
Basophils (%)	0–3
Platelets (× 10^3/mm^3)	250–850

Biochemical data

Serum Protein (g/100 ml)	4.6–6.2
Albumin (g/100 ml)	2.1–3.9
Globulin (g/100 ml)	1.7–2.6
Glucose (mg/100 ml)	60–125
Blood urea nitrogen (mg/100 ml)	9–31.5
Creatinine (mg/100 ml)	0.6–2.2
Total bilirubin (mg/100 ml)	0.3–0.9
Cholesterol (mg/100 ml)	20–43

GERBIL

Biological data

Adult weight (g)	Male 65–100 Female 55–85
Diploid number	44
Food intake	5–8 g/day
Water intake	4–7 ml/day
Lifespan (years)	3–4
Rectal temperature (°C)	37–38.5
Heart rate/min	360
Blood volume (ml/kg)	66–78
Respiratory rate/min	90

Breeding data

Puberty	Vaginal opening 41 days (28 g)
Age to breed male (days)	70–85
Age to breed female (days)	65–85
Gestation (days)	24–26 (27–48 if lactating)
Litter size	3–7
Birth weight (g)	2.5–3.5 (depends on litter size)
Weaning age	21 days
Oestrous cycle	4–5 days
Post-partum oestrus	Fertile

Haematological data

RBC ($\times 10^6$/mm^3)	8–9*
PCV (%)	43–49
Hb (g/100 ml)	12.6–16.2
WBC ($\times 10^3$/mm^3)	7–15
Neutrophils (%)	5–34
Lymphocytes (%)	60–95
Eosinophils (%)	0–4
Monocytes (%)	0–3
Basophils (%)	0–1
Platelets ($\times 10^3$/mm^3)	400–600

Biochemical data

Serum Protein (g/100 ml)	4.3–12.5
Albumin (g/100 ml)	1.8–5.5
Globulin (g/100 ml)	1.2–6
Glucose (mg/100 ml)	50–135
Blood urea nitrogen (mg/100 ml)	17–27
Creatinine (mg/100 ml)	0.6–1.4
Total bilirubin (mg/100 ml)	0.2–0.6
Cholesterol (mg/100 ml)	90–150

* High levels of reticulocytes, stippled red blood cells, and polychromatic cells.

HAMSTER
Biological data

Adult weight (g)	Male 85–130
	Female 95–150
Diploid number	44
Food intake	>15 g/100 g bw
Water intake	>10 ml/ 100 g bw
Lifespan (years)	1.5–3
Rectal temperature (°C)	37–38
Heart rate/min	250–500
Blood pressure systole (mmHg)	150
Blood pressure diastole (mmHg)	100
Blood volume (ml/kg)	78
Respiratory rate/min	35–135
Tidal volume (ml)	0.6–1.4

Breeding data

Puberty (days)	28–42
Age to breed male (weeks)	10–14
Age to breed female (weeks)	6–12
Gestation (days)	15–16
Litter size	5–9
Birth weight (g)	2
Weaning age (days)	20–25
Oestrous cycle (days)	4
Post-partum oestrus	Infertile

Haematological data

RBC (×10^6/mm^3)	6–10
PCV (%)	36–55
Hb (g/100 ml)	10–16
WBC (×10^3/mm^3)	3–11
Neutrophils (%)	10–42
Lymphocytes (%)	50–95
Eosinophils (%)	0–4.5
Monocytes (%)	0–3
Basophils (%)	0–1
Platelets (×10^3/mm^3)	200–500

Biochemical data

Serum Protein (g/100 ml)	4.5–7.5
Albumin (g/100 ml)	2.6–4.1
Globulin (g/100 ml)	2.7–4.2
Glucose (mg/100 ml)	60–150
Blood urea nitrogen (mg/100 ml)	12–25
Creatinine (mg/100 ml)	0.91–0.99
Total bilirubin (mg/100 ml)	0.25–0.6
Cholesterol (mg/100 ml)	25–135

RABBIT
Biological data

Adult weight (g)	900–6000
Diploid number	44
Food intake	5 g/100 g bw
Water intake	10 ml/100 g bw
Lifespan (years)	6–12
Rectal temperature (°C)	38.5–40
Heart rate/min	130–325
Blood pressure systole (mmHg)	90–130
Blood pressure diastole (mmHg)	60–90
Blood volume (ml/kg)	57–65
Respiratory rate/min	30–60
Tidal volume (ml)	4–6

Breeding data

Puberty (days)	90–120
Age to breed male (months)	6–10
Age to breed female (months)	4–9
Gestation (days)	30–32
Litter size	4–10
Birth weight (g)	30–70
Weaning age (weeks)	4–8
Oestrous cycle	Induced ovulator
Post-partum oestrus	None (not used)

Haematological data

RBC ($\times 10^6$/mm³)	4–7
PCV (%)	36–48
Hb (g/100 ml)	10–15.5
WBC ($\times 10^3$/mm³)	9–11
Neutrophils (%)	20–75*
Lymphocytes (%)	30–85
Eosinophils (%)	0–4
Monocytes (%)	1–4
Basophils (%)	2–7
Platelets ($\times 10^3$/mm³)	250–270

Biochemical data

Serum Protein (g/100 ml)	5.4–7.5
Albumin (g/100 ml)	2.7–4.6
Globulin (g/100 ml)	1.5–2.8
Glucose (mg/100 ml)	75–150
Blood urea nitrogen (mg/100 ml)	17–23.5
Creatinine (mg/100 ml)	0.8–1.8
Total bilirubin (mg/100 ml)	0.25–0.74
Cholesterol (mg/100 ml)	35–53

* Neutrophils often resemble eosinophils due to cytoplasmic granules.

DOG (BEAGLE)
Biological data

Adult weight (kg)	10–15
Diploid number	78
Daily food intake	18 g/kg bw of complete dry diet
Daily water intake	1200 ml*
Lifespan (years)	12
Rectal temperature (°C)	37.9–39.9
Heart rate/min	70–160
Blood pressure systole (mmHg)	95–136
Blood pressure diastole (mmHg)	43–66
Blood volume (ml/kg)	76–107
Respiratory rate/min	22
Tidal volume (ml)	251–432

Breeding data

Puberty (months)	6–9
Age to breed male (years)	1–2
Age to breed female (years)	1–2
Gestation (days)	59–68 (average 63)
Litter size	1–12 (usually 4–6)
Birth weight (g)	250
Weaning age (weeks)	4–6
Oestrous cycle	Monoestrous. 2 oestrous periods per year
Post-partum oestrus	No

Haematological data

RBC (×10⁶/mm³)	5.5–8.5
PCV (%)	37–55
Hb (g/100 ml)	12–18
WBC (×10³/mm³)	6–17
Neutrophils (%)	60–70
Lymphocytes (%)	12–30
Eosinophils (%)	2–10
Monocytes (%)	3–10
Basophils (%)	Rare
Platelets (×10³/mm³)	200–900

Biochemical data

Serum Protein (g/100 ml)	6–7.5
Albumin (g/100 ml)	3–4
Globulin (g/100 ml)	2.4–3.7
Glucose (mg/100 ml)	54–99
Blood urea nitrogen (mmol/l)	3.5–7.5
Creatinine (μmol/l)	<120
Total bilirubin (μmol/l)	<5.0
Cholesterol (mmol/l)	4–7

* Variation with size and diet composition.

CAT

Biological data		Breeding data	
Adult weight (kg)	2.5–3.5	Puberty (months)	Male 8–9 Female 5–12
Diploid number	38	Age to breed male (years)	1
Daily Food intake (g)	200	Age to breed female (years)	1
Daily water intake (ml)	500*	Gestation (days)	58–65 (average 63)
Lifespan (years)	12–18	Litter size	3–5
Rectal temperature (°C)	38.1–39.2	Birth weight (g)	110–120
Heart rate/min	110–140	Weaning age (weeks)	4–6
Blood pressure systole (mmHg)	120	Oestrous cycle	Seasonal poly-oestrus.
Blood pressure diastole (mmHg)	75		Cycles 2–3 weeks long
Blood volume (ml/kg)	47–65		January to September.
Respiratory rate/min	26		Induced ovulator
Tidal volume (ml)	34	Post-partum oestrus	No

Haematological data		Biochemical data	
RBC ($\times 10^6$/mm^3)	5–10	Serum Protein (g/100 ml)	6–7.5
PCV (%)	30–45	Albumin (g/100 ml)	2.5–4.0
Hb (g/100 ml)	8–15	Globulin (g/100 ml)	2.5–3.8
WBC ($\times 10^3$/mm^3)	5.5–19.5	Glucose (mg/100 ml)	81–108
Neutrophils (%)	35–75	Blood urea nitrogen (mmol/l)	3.5–8.0
Lymphocytes (%)	20–55	Creatinine (μmol/l)	<180
Eosinophils (%)	2–12	Total bilirubin (μ mol/l)	<4
Monocytes (%)	1–4	Cholesterol (mmol/l)	2–4
Basophils (%)	Rare		
Platelets ($\times 10^3$/mm^3)	300–700		

* Depends on composition of diet.

FERRET

Biological data

Adult weight (g)	600–2000[a]
Food intake (g)	50–75/day
Water intake (ml)	75–100/day[b]
Lifespan (years)	5–9
Rectal temperature (°C)	39
Heart rate/min	250
Blood pressure systole (mmHg)	110–140
Blood pressure diastole (mmHg)	31–35
Respiratory rate/min	33–36

Breeding data

Puberty (months)	9–12
Age to breed male (days)	365
Age to breed female (days)	275
Gestation (days)	38–44 (average 42)
Litter size	8
Birth weight (g)	7–10
Weaning age (weeks)	6–7
Oestrous cycle	Induced ovulator

Haematological data

RBC ($\times 10^6$/mm^3)	6.8–12.2
PCV (%)	42–61
Hb (g/100 ml)	15–18
WBC ($\times 10^3$/mm^3)	4–19
Neutrophils (%)	11–84
Lymphocytes (%)	12–54
Eosinophils (%)	0–7
Monocytes (%)	0–9
Basophils (%)	0–2
Platelets ($\times 10^3$/mm^3)	297–910

Biochemical data

Serum Protein (g/100 ml)	5.1–7.4
Albumin (g/100 ml)	2.6–3.8
Globulin (g/100 ml)	2.5–4.8
Glucose (mg/100 ml)	94–207
Blood urea nitrogen (mg/100 ml)	10–45
Creatinine (mg/100 ml)	0.4–0.9
Total bilirubin (mg/100 ml)	<1
Cholesterol (mg/100 ml)	64–296

[a] Both sexes show a periodic weight fluctuation of 30–40 per cent. Fat is laid down in autumn and lost in winter.
[b] Dry carnivore pelleted diet. This should be fed soaked in hot water to form a paste.

MARMOSET

Biological data		Breeding data	
Adult weight (g)	300–350	Puberty (months)	8
Diploid number	46	Age to breed (years)	1.5–2
Food intake	20 g daily (New World monkey pelleted diet)	Gestation (days)	140–148 (average 145)
Water intake	*Ad libitum*	Litter size	Usually dizygotic twins (can be 1–4 offspring)
Lifespan (years)	10–16	Birth Weight (g)	25–35
Blood volume (ml/kg)	70	Weaning age	6 weeks to 6 months*
		Oestrous cycle (days)	14–18. Few overt signs of oestrus.
		Comments	Interbirth interval 154–178 days

Haematological data		Biochemical data	
RBC ($\times 10^6$mm³)	5.7–6.95	Serum protein (g/100 ml)	6.6–7.1
PCV (%)	42–52	Albumin (g/100 ml)	3.8
Hb (g/100 ml)	14.9–17	Globulin (g/100 ml)	2.7–3.9
WBC ($\times 10^3$/mm³)	7.3–12.8	Glucose (mg/100 ml)	126–228
Neutrophils (%)	26–62	Blood urea nitrogen (mg/100 ml)	51.8
Lymphocytes (%)	30–67	Creatinine (mg/100 ml)	0.9–1.2
Eosinophils (%)	0.6–4.2	Total bilirubin (mg/100 ml)	0.4–0.6
Monocytes (%)	0.4–5		
Basophils (%)	0.1–1.1		
Platelets ($\times 10^3$/mm³)	490		

* Young can be suckled for longer, particularly if stressed.

SQUIRREL MONKEY

Biological data		Breeding data	
Adult weight (g)	Male 550–1100	Puberty (years)	Male 3.5
	Female 350–750		Female 1.5
Diploid number	44	Age to breed Male (years)	3.5
		Age to breed female (years)	2.5–3.5
Food intake (g)	45–60 daily	Gestation	145–150 days
Water intake (ml/kg)	100–300 daily 500 if infant	Litter size	1
Lifespan (years)	10	Birth Weight (g)	100
Rectal temperature (°C)	39.7	Oestrous cycle	7–13 days
Heart rate/min	215–265	Comments	Seasonal polyoestrus. Mating March to May
Blood pressure systole (mmHg)	127–160		
Blood pressure diastole (mmHg)	77–83		
Respiratory rate/min	55–70		
Tidal volume (ml)	7.5–8.9		

Haematological data		Biochemical data	
RBC ($\times 10^6$mm^3)	7.5	Serum protein (g/100 ml)	6.6–7.8
PCV (%)	42	Albumin (g/100 ml)	4.9–6.3
Hb (g/100 ml)	14.1	Globulin (g/100 ml)	3.16
WBC ($\times 10^3$/mm^3)	8.0	Glucose (mg/100 ml)	57–91
Neutrophils (%)	51	Blood urea nitrogen (mg/100 ml)	25.6–38.4
Lymphocytes (%)	41	Total bilirubin (mg/100 ml)	0.2
Eosinophils (%)	5	Cholesterol (mg/100 ml)	199.1
Monocytes (%)	3		
Basophils (%)	<1		

MACAQUES

Biological data	Rhesus macaque (*Macaca mulatta*)	Cynomolgus macaque (*Macaca fascicularis*)
Adult weight (kg)	Male 6–11, Female 4–9	Male 4–8, Female 2–6
Diploid number	42	42
Food intake	420 J/kg for maintenance, 525–630 J/kg for production, 840 J/kg for neonates	
Water intake	*Ad libitum.* 1180 ml/m² body surface area required daily	
Lifespan (years)	20–30	15–25
Temperature (°C)	36–40	37–40
Heart rate/min	120–180*	240
Blood pressure systole (mmHg)	125	As rhesus
Blood pressure diastole (mmHg)	75	As rhesus
Blood volume (ml/kg)	55–80	50–96
Respiratory rate/min	32–50*	

* Values determined under sedation.

Breeding Data

Age at puberty (years)	Male 3–4, Female 2–3	3–4
Age to breed (years)	Male 4–5, Female 3–5	4–5
Gestation (days)	146–180 (average 164)	153–179 (average 167)
Litter Size	1	1
Birth weight (kg)	0.4–0.55	0.33–0.35
Weaning age (months)	7–14 Can hand rear from birth	14–18
Breeding cycle	Menstrual cycle 28 days	Menstrual cycle 31 days
Comments	Seasonal breeding September to January	Non-seasonal breeding

Haematological data

RBC (×10⁶mm³)	3.56–6.95
PCV (%)	26–48
Hb (g/100 ml)	8.8–16.5
WBC (×10³/mm³)	2.5–26.7
Neutrophils (%)	5–88
Lymphocytes (%)	8–92
Eosinophils (%)	0–14
Monocytes (%)	0–11
Basophils (%)	0–6
Platelets (×10³/mm³)	109–597

Biochemical data

Serum protein (g/100 ml)	4.9–9.3
Albumin (g/100 ml)	2.8–5.2
Globulin (g/100 ml)	1.2–5.8
Glucose (mg/100 ml)	46–178
Blood urea nitrogen (mg/100 ml)	8–40
Creatinine (mg/100 ml)	0.1–2.8
Total bilirubin (mg/100 ml)	0.1–2
Cholesterol (mg/100 ml)	108–263

BABOON

Biological data

Adult weight (kg)	Male 22–30 Female 11–15
Diploid number	42
Lifespan (years)	Up to 28
Rectal temperature (°C)	36–39
Heart rate/min	74–200
Blood pressure systole (mmHg)	135
Blood pressure diastole (mmHg)	80
Blood volume (ml/kg)	62–65
Respiratory rate/min	29

Breeding data

Puberty (years)	2.5–3
Age to breed male (years)	4.5–5
Age to breed female (years)	3.5–4
Gestation (days)	164–186 (average 170)
Litter Size	1
Birth Weight (kg)	0.87–0.94
Weaning age (months)	5–8
Breeding cycle	Menstrual cycle 36 days
Comments	Non-seasonal breeding

Haematological data

RBC ($\times 10^6$mm^3)	4.8
PCV (%)	35–40
Hb (g/100 ml)	11.9–12.7
WBC ($\times 10^3$/mm^3)	7.5–9.6
Neutrophils (%)	51
Lymphocytes (%)	43
Eosinophils (%)	3
Monocytes (%)	2–8
Basophils (%)	0.2

Biochemical data

Serum protein (g/100 ml)	6.6
Albumin (g/100 ml)	3.8
Globulin (g/100 ml)	2.8
Glucose (mg/100 ml)	95.9
Blood urea nitrogen (mg/100 ml)	12
Creatinine (mg/100 ml)	1.28
Total bilirubin (mg/100 ml)	0.33

SHEEP

Biological data

Adult weight (kg)	70 (much breed variation)
Diploid number	54
Daily food intake (kJ)	12.1
Daily water intake (l)	3.6 (more if lactating)
Lifespan (years)	10–15
Rectal temperature (°C)	39.5
Heart rate/min	60–120
Blood volume (ml/kg)	58–64
Respiratory rate/min	12–20

Breeding data

Puberty (months)	7–12
Age to breed male	First or second autumn
Age to breed female	First or second autumn
Gestation (days)	140–160 (average 150)
Litter size	2
Birth weight (kg)	3.5
Weaning age (weeks)	16–20
Oestrous cycle	Seasonal poly-oestrus September to March. Cycle 12–19 days (average 17)
Post-partum oestrus	No

Haematological data

RBC ($\times 10^6$mm³)	8–13
PCV (%)	24–40
Hb (g/100 ml)	8–16
WBC ($\times 10^3$/mm³)	4–12
Neutrophils (%)	20–40
Lymphocytes (%)	40–70
Eosinophils (%)	0–15
Monocytes (%)	1–12
Basophils (%)	0–1

Biochemical data

Serum protein (g/100 ml)	6.6–8.1
Albumin (g/100 ml)	2.4–3
Globulin (g/100 ml)	3.5–5.7
Glucose (mg/100 ml)	100–500
Blood urea nitrogen (mg/100 ml)	8–20
Creatinine (mg/100 ml)	1.2–1.9
Total bilirubin (mg/100 ml)	0.1–0.42
Cholesterol (mg/100 ml)	52–76

GOAT

Biological data			Breeding data	
Adult weight (kg)	70 (much breed variation)		Puberty (months)	3–8 (average 7)
Diploid number	60		Age to breed female	Second autumn after birth
Food intake (kJ)	12.6		Gestation (days)	147–155 (average 151)
Water intake (l)	3.5 (more if lactating)		Litter size	1–3
Lifespan (years)	10–15		Birth weight (kg)	2–3
Rectal temperature (°C)	38–40		Oestrous Cycle	Seasonal poly-oestrus September to January. Cycle 18–21 days.
Heart rate/min	70–135		Post-partum oestrus	No
Blood volume (ml/kg)	57–90			
Respiratory rate/min	15–23			

Haematological data		Biochemical data	
RBC ($\times 10^6$mm^3)	12–14	Serum protein (g/100 ml)	5.4–7.5
PCV (%)	24–48	Albumin (g/100 ml)	2.7–4.6
Hb (g/100 ml)	8–14	Globulin (g/100 ml)	1.5–2.8
WBC ($\times 10^3$/mm^3)	5–14	Glucose (mg/100 ml)	75–150
Neutrophils (%)	20–40	Blood urea nitrogen (mg/100 ml)	17–23.5
Lymphocytes (%)	50–65	Creatinine (mg/100 ml)	0.8–1.8
Eosinophils (%)	3–8	Total bilirubin (mg/100 ml)	0.25–0.74
Monocytes (%)	1–5	Cholesterol (mg/100 ml)	35–53
Basophils (%)	0–1		

PIG

Biological data	Large White	Yucatan
Adult weight (kg)	220–250	70–80
Food intake	Needs vary with production status	
Water intake	2–3 times food intake	
Lifespan (years)	16–18	8
Temperature (°C)	38.7–39.7	38.7–39.7
Heart rate/min	50–100	50–100
Blood pressure systole (mmHg)	150	150
Blood volume (ml/kg)	56–69	56–69
Respiratory rate/min	10–16	10–16

Breeding data

Age at puberty (months)	4–9 (average 7)	4
Age to breed	Male 7 months Female after third oestrus	Male 5 months Female 5 months
Gestation (days)	110–116 (average 113)	113–114
Litter size	8–12	6
Birth weight (kg)	1.3	0.56
Weaning age	3 weeks	3 weeks
Oestrous cycle	14–26 days (average 21)	19.5 days
Post-partum oestrus	Yes. Not used, and non-ovulatory	

Haematological data

RBC ($\times 10^6$mm³)	5–7	5–10
PCV (%)	32–45	22–40
Hb (g/100 ml)	10–16	as Large White
WBC ($\times 10^3$/mm³)	7–20	4–10
Neutrophils (%)	30–50	as Large White
Lymphocytes (%)	40–60	as Large White
Eosinophils (%)	0–10	as Large White
Monocytes (%)	2–10	as Large White
Basophils (%)	0–1	as Large White
Platelets ($\times 10^3$/mm³)	350–700	as Large White

PIGS
Biochemical data (for Yucatan and Large White)

Serum protein (g/100 ml)	6–8.9
Albumin (g/100 ml)	1.6–3.8
Globulin (g/100 ml)	5.2–6.4
Glucose (mg/100 ml)	100–500
Blood urea nitrogen (ml/100 ml)	10–30
Creatinine (mg/100 ml)	1–2.7
Total Bilirubin (mg/100 ml)	0–0.6
Cholesterol (mg/100 ml)	36–54

CATTLE

Biological data		Breeding data	
Adult weight (kg)	Male 660–1000 Female 400–800	Puberty (months)	12–15
Food intake (kg)	Depends on reproductive status. <90 kg grass, <70 kg silage, and concentrates	Age to breed (years)	1–2
Water intake (ml)	*Ad libitum*	Gestation (days)	274–291 (average 282)
Lifespan (years)	15–20	Litter size	1–2
Rectal temperature (°C)	38–39	Birth weight (kg)	25–45
Heart rate/min	40–100	Weaning age (months)	6–8*
Blood volume (ml/kg)	57–62	Oestrous cycle (days)	18–24
Respiratory rate/min	27–40	Post-partum oestrus	Yes

Haematological data		Biochemical data	
RBC ($\times 10^6$mm^3)	5–10	Serum protein (g/100 ml)	5.3–7.5
PCV (%)	24–46	Albumin (g/100 ml)	2.1–3.6
Hb (g/100 ml)	8–15	Globulin (g/100 ml)	3–5.5
WBC ($\times 10^3$/mm^3)	4–12	Glucose (mmol/l)	2–3.2
Neutrophils (%)	6–45	Blood urea nitrogen (mmol/l)	2–6.6
Lymphocytes (%)	18–75	Creatinine (μmol/l)	44–165
Eosinophils (%)	2	Total bilirubin (μ mol/l)	0–6.5
Monocytes (%)	1–7	Cholesterol (mmol/l)	1–3
Basophils (%)	Rare		
Platelets ($\times 10^3$/mm^3)	100–800		

* Most dairy calves are removed soon after birth and hand-reared.

HORSE

Biological data		Breeding data	
Adult weight (kg)	Up to 500. Much breed variation	Puberty (months)	12–18
Food intake	Depends on work and breed	Age to breed (years)	2–3 (depends on use)
Water intake	*Ad libitum*	Gestation (days)	321–362 (average 336)
Lifespan (years)	<30	Litter size	1 (twins do not survive)
Rectal temperature (°C)	37.6–38.2	Birth weight (kg)	Depends on breed
Heart rate/min	23–70	Weaning age (months)	6
Blood volume (ml/kg)	75	Oestrous cycle	Seasonal poly-oestrus, May–November. Cycle 13–25 days (average 21)
Respiratory rate/min	8–12	Post-partum oestrus	Yes (not used)

Haematological data		Biochemical data	
RBC ($\times 10^6$mm^3)	7–14[a]	Serum protein (g/100 ml)	6–7.3
PCV (%)	29–47	Albumin (g/100 ml)	2.5–3.8
Hb (g/100 ml)	10–16.9	Globulin (g/100 ml)	3–4.8
WBC ($\times 10^3$/mm^3)	4.1–10.1	Glucose (mmol/l)	2.5–5.5
Neutrophils (%)	14–85	Blood urea nitrogen (mmol/l)	2.5–7
Lymphocytes (%)	14–77	Creatinine (μmol/l)	50–147
Eosinophils (%)	0–7	Total bilirubin (μmol/l)	17–34[b]
Monocytes (%)	0–2	Cholesterol (mmol/l)	2.3–3
Basophils (%)	Rare		
Platelets ($\times 10^3$/mm^3)	120–360		

[a] Depends whether hot-blooded or cold-blooded
[b] Horses have no gall bladder, hence high circulating bilirubin

9 Handling of laboratory species

Good animal handling techniques will reduce the risk of injury from bites and scratches and will increase the confidence of both the handler and the animal, thus reducing stress to all those involved.

Stress will induce biochemical and physiological responses in the animal which can effect experimental results. All animals will respond in some way to the presence of a human and most species can recognize individuals and will be nervous of strangers. It is therefore important for the licensee to establish a friendly relationship with the animal to reduce nervousness on both sides. An animal that is confident and relaxed with its handler will be more co-operative enabling procedures to be carried out more easily.

Some aspects of handling will vary according to the species. A brief description of techniques for each species follows, but with each it is essential to get advice and assistance from people with previous experience such as the day-to-day care person.

MOUSE

Mice move very fast, so you have to be quick and decisive to catch them.

1. Grasp the *base* of the tail gently but firmly and lift the mouse. Mice can be transferred between boxes this way.
2. Place the mouse down on a non-slip surface (e.g. top of cage), without releasing the tail.
3. Grasp the scruff between the thumb and index finger of the other hand.
4. Hold the tail down with the 4th and 5th finger for extra restraint (see Figure 9.1).

Sexing: Ano-genital distance is greater in male than female (approximately twice).

To handle newborn mice, first transfer the mother to a separate cage to prevent any aggression from her. The newborn mice can then be rolled into the palm. Handle as adults from 10 days, but still remove the mother first. To avoid subsequent cannibalism by the mother, rub the young with nest material before replacing to mask the human smell, or handle the mother thoroughly so that she will also acquire a human smell.

9.1 Handling the mouse.

RAT

Rats should not be grasped by the tail, except at the base and only for short periods. Adults may be restrained by placing the 2nd and 3rd fingers either side of the mandible, with the thumb and 4th and 5th fingers either side of the chest.

Alternatively, the rat may be restrained by holding the base of the tail with one hand and sliding the other up the body to a position behind the shoulders. Hold the thorax between thumb and fingers. (see Figure 9.2).

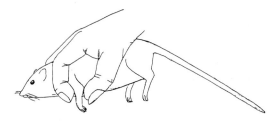

9.2 Handling the rat.

GUINEA-PIG

Guinea-pigs can be lifted using one hand over the shoulders as in the rat, and supporting the body with the other hand. The guinea-pig can then be turned over for intraperitoneal injections or sexing.

Alternatively, one hand may be placed under the thorax and the other under the rear feet. It is particularly important to support pregnant females with two hands. (see Figure 9.3).

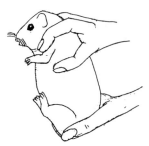

9.3 Handling the guinea-pig.

GERBIL AND HAMSTER

Gerbils may be restrained using the same techniques as for mouse and rat. They may be picked up by grasping the base of the tail, then supporting the body with the other hand. Great care must be taken because if the shaft of the tail is held the skin may slip off.

Hamsters have a tendency to bite if startled and should be picked up gently but firmly. Hamsters have no tail, and large cheek pouches. They are restrained by grasping a **large** pinch of scruff and turning over into the hand as for mice. If insufficient scruff is grasped, the hamster will turn and bite. Additional restraint may be achieved by grasping skin along the back between the fingers and palm.

RABBIT

The ears should **NOT** be used for lifting a rabbit. The scruff is grasped in one hand, and the other hand placed behind the rear feet. The rabbit can then be lifted. It may be carried cradled in the arms with the head and eyes hidden under the upper arm.

It is particularly important to restrain rabbits correctly, as struggling can result in injuries to the spine. Restraint can be achieved by using a specially designed rabbit box, or by wrapping the body and legs of the animal firmly in a towel such that only the head and ears protrude.

To sex adult rabbits, the head is grasped with the thumb and third finger each side of the head and the index finger between the ears. This way the head is controlled. The other hand can be placed under the abdomen, and the rabbit turned over on to the knee, its back being supported by the forearm. The other hand can be used to assist in sexing the rabbit (See Figure 9.4).

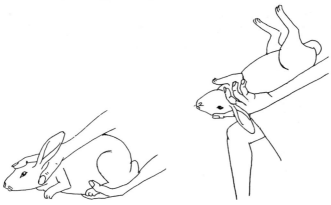

9.4 Handling the rabbit.

DOG

When approaching dogs, the back of a hand should first be offered for the animal to sniff. An open hand may be interpreted as a threat and be bitten. The lead and collar, or the scruff of the neck if there is no lead, can be grasped in the other hand once the dog has accepted the person.

To lift a dog, approach the dog from the left-hand side. Restrain the head with the left hand by holding its collar or scruff (gently), and lift the dog by placing the right arm over the body of the dog and taking its weight on its sternum (See Figure 9.5).

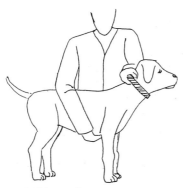

9.5 Lifting a dog.

If a dog is nervous or aggressive, it may be muzzled, using a ready-made plastic or fabric muzzle, or with a tape (see Figure 9.6).

9.6 A tape muzzle on a dog.

For venepuncture, an assistant places their left arm under the dog's neck and holds its head firmly against their chest. The right arm is then placed over the thorax, and the right elbow is used to hold the dog's chest against the person's chest. The first two fingers of the right hand are placed behind the dog's elbow to hold the forearm out, and the thumb is placed firmly over the top of the leg to raise the vein. Good results are obtained if the thumb is first placed towards the inside of the leg then rotated outwards slightly, as this brings the vein on to the top of the leg (see Figure 9.7). The quick release tourniquet can be used (see Figure 10.1a).

These methods may be modified for facilitating the administration of medicines.

9.7 Holding a dog for venepuncture.

CAT

Cats are dignified creatures which respond to gentle handling and a quiet approach. A cat should first be allowed to familiarize itself with the smell of the handler by sniffing the back of the hand, as for dogs. Most cats can then be picked up by placing one hand under the thorax and the other behind the rear end.

Difficult cats can be restrained by grasping the scruff. Care must be taken though as the cat will still be able to reach the handler with its claws. Particularly vicious cats may be wrapped in a towel or blanket, or placed in a crush cage and given a sedative.

Generally, cats are amenable and can be held for minor procedures and examinations by holding them gently but firmly around the shoulders (see Figure 9.8).

For venepuncture, the method described for the dog can be used.

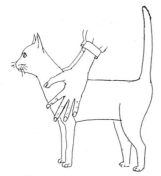

9.8 Restraining a cat.

FERRET

Although friendly, ferrets will bite if startled. They should be picked up using an over the shoulder grip, with the thumb one side of the mandible and the fingers on the other (see Figure 9.9).

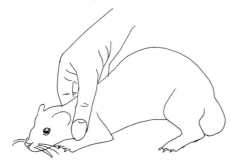

9.9 Handling a ferret.

PRIMATES

Particular care must be taken when handling primates because of the danger of transmitting potentially zoonotic diseases. Appropriate protective clothing must be worn. For quarantine animals, this should include cap, gown, mask, boots, and gloves. For other animals, gloves and protective gowns may be sufficient but there will be local safety rules for each institution which must be followed. These rules will take into account the origin of the animal and the results of health screening (see Chapter 7).

All but the smallest primates should be sedated prior to handling. Small New World monkeys are first caught using leather gloves, and sedated with alphaxalone/alphadolone (Saffan, Pitman Moore).

For larger monkeys, e.g. Old World monkeys, cages usually have crush backs or push fronts. The animals are held against the front or rear of the cage and an injection is given through the bars. Ketamine is used to sedate Old World monkeys.

Once sedated, primates should be held by the upper arms, held horizontally behind the shoulders, to keep their faces away from the handler (see Figure 9.10).

9.10 Holding a primate.

PIG

Pigs are easily startled and so should be approached patiently and quietly. Any attempts at restraining or handling pigs are usually accompanied by loud squealing, which ceases only when the hold is released.

Groups of pigs can be moved around using boards, as pigs will turn away from solid surfaces (see Figure 9.11).

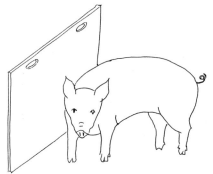

9.11 Use of a board to move a pig.

Small piglets may be picked up by one hind leg, held above the hock. Larger piglets can be tucked under one arm.

Adult pigs can be restrained by using a rope snare placed around the upper jaw, caudal to the canines. The pig pulls away from this, causing the rope to tighten over the nose. The rope can then be tied to a fence or pole (see Figure 9.12).

9.12 Use of a rope to restrain a pig.

SHEEP AND GOAT

Both sheep and goats will usually run away when an attempt is made to catch them, but sheep are particularly easily startled. Patience and gentle handling are required for success.

Catching sheep by grasping the fleece may well result in handfuls of wool being pulled out! Sheep can be restrained by placing one hand under the chin and the other behind the head. This is usually sufficient restraint for minor examinations and procedures.

To cast a sheep, the handler stands on the left side of the sheep, grasps the muzzle with the left hand, and pushes the head round to face the tail. This can be facilitated by placing the thumb in the diastema behind the incisor teeth. The sheep can then be pulled over towards the handler (see Figure 9.13).

Alternatively, once caught, the sheep can be grasped around the sternum with the right hand and lifted up so that it comes to rest sitting on its rump.

Goats are usually horned, which can be dangerous for the handler but means that there is a convenient handle to get hold of. Goats are less nervous than sheep and can be caught quite easily with a little patience. Goats without horns can be caught like sheep.

Casting of goats is rarely required.

9.13 Casting a sheep.

10 Procedural data

It is likely that when animals are used for scientific purposes, samples will need to be taken from them or substances will be administered by one route or another. These procedures will in themselves have some effects on the animal, and it is essential to minimize any distress caused. This chapter gives **guidelines** on safe volumes and sites for administration and withdrawal of substances, but the exact details of the procedure will be stated on the relevant project licence, and **this must be checked prior to carrying out any procedure**. The licensee must also have authorization to perform the procedure on his or her personal licence.

REMOVAL OF BLOOD

The removal of blood from an animal is a procedure with three potentially stressful components.

1. Handling and restraining the animal is stressful. To minimize the distress caused, the licensee should be familiar with humane methods of handling and restraint (see Chapter 9), and should consider using an appropriate sedative or anaesthetic (see Chapter 12). Many animals can be trained to accept the handling required to take blood samples, such as cats, dogs, rabbits, and pigs, and although this takes time, it is worthwhile.

2. Venepuncture causes some pain and discomfort, whatever the site, and requires considerable skill. The expertise required to carry this out successfully must be gained first by watching others, then by practising on cadavers or models such as the KOKEN rat, and then by carrying out the technique oneself, once a licence has been granted, under direct supervision.

3. The removal of blood causes physiological responses, the magnitude of which depend on the volume of blood removed

(as a percentage of the total), and the speed of withdrawal. The rapid removal of large quantities of blood will cause the animal to go into hypovolaemic shock, and may even cause death. The percentage blood loss required to cause hypovolaemic shock varies with the speed of withdrawal, whether or not fluid is replaced concurrently, and the psychophysiological state of the animal at the time. Chronic slow haemorrhage is tolerated better than acute blood loss, and placid animals tolerate greater losses than nervous ones, again indicating the need for competent handling and training of the animals.

Stress responses in the animal result in the release of hormones and other substances to counteract the stress, which can cause anomalous experimental results. Therefore, it is essential to minimize the stress caused to an animal when removing blood from the humane viewpoint, and also to ensure good scientific practice.

It is important to refine the experimental technique such that the quantity of blood removed is minimized, to reduce the stresses placed on the animal. This is particularly important in small mammals, such as mice, where the blood volume is small and sample volume is critical.

The withdrawal of blood from any vessel requires skill in handling the animal and in manipulating the equipment. The person taking the samples should be fully familiar with the chosen technique, and have all the equipment ready before starting.

Quality of samples

To achieve meaningful results and avoid needless repetition, any samples taken must be of good quality, and be preserved in the best possible manner. If the sampling technique is poor, blood may clot or haemolyse rendering results invalid. To avoid these problems, samples must be taken skilfully, and treated appropriately thereafter.

Blood may be collected using syringes and hypodermic or butterfly needles, through indwelling cannulae, with double ended needles and evacuated tubes (e.g. Vacutainers), or in very small species by incision of a vein using a sterile lancet or scalpel blade. The latter is not recommended however, as inadvertent movement can result in injury. If needles are used, the needle should be as large as is practicable for the species. This allows blood to flow faster, reducing the likelihood of clotting, and also causes less damage to the red cells, reducing the possibility of haemolysis.

Thought should be given to the desirability of using anticoagulants. Different anticlotting agents are suitable for different purposes:

No anticoagulant: Blood clots, and serum can be removed after centrifugation.

Lithium heparin: This is the anticoagulant of choice for most biochemical assays. The yield of plasma from heparinized blood is greater than the yield of serum from clotted blood, which may make heparin a good choice for collecting blood for harvesting antibodies.

Potassium EDTA: (ethylene diaminetetra-acetate). This is used for haematological analyses.

Oxalate/fluoride: This is used for blood glucose determination.

Several other anticoagulants are available, e.g. for collecting blood for transfusions or for analysis of clotting factors.

After collection into anticoagulant, the blood should be mixed thoroughly by **rolling**, not by shaking as this can damage the cells and lead to haemolysis.

Analysis of samples

It is preferable for samples to be submitted fresh for analysis. If this is not possible, samples may need to be refrigerated or deep frozen. For some enzyme determinations, degeneration of the enzyme renders analysis useless if performed more than a few hours after blood collection. It is advisable to determine the exact requirements of the laboratory protocol prior to sample collection.

Restraint

When taking samples, the animal should be restrained by an experienced handler who is known to the animal. This markedly reduces the stress placed upon the animal. Chemical restraint is generally only required in primates, or if the technique involves more than a pinprick, e.g. for tail tip amputation (BVA/FRAME/RSPCA/UFAW 1993). The handler may also be required to raise the vein, by occluding the venous drainage proximal to the site of venepuncture. This must be performed correctly, or withdrawal of blood will be difficult. The use of a quick release tourniquet, such as a Vetourniquet (Vet-2-Vet), greatly facilities blood sampling in some species and vasodilating agents may be used in other species (e.g. Vasolate, International Market Supply—IMS). (See Figure 10.1 a and b.)

Site and location of the vein

It is important to be certain of the location of the vein, either by visualizing it or palpating its course, and to have it immobilized, before piercing the skin. If unsure of its position, venepuncture should not be attempted. If it has been correctly raised, locating the vein will be facilitated.

Preparation of the site

Blood should be collected using an aseptic technique. The area should be clipped to remove hair, then cleaned. The use of warm water with or without disinfectant will help dilate superficial veins as well as cleansing the skin. After cleansing, the skin can be

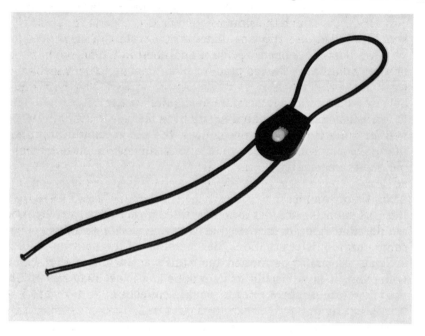

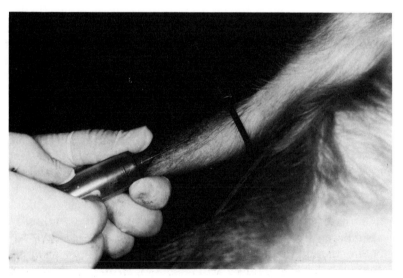

10.1 (a) Quick release tourniquet. (b) The tourniquet in use.

swabbed with 70% ethanol or disinfectant. In some species, it may be advantageous to apply local anaesthetic cream (e.g. EMLA, Astra) to the site 30–60 minutes before venepuncture to prevent any discomfort in such animals as the cat, dog, rabbit or pig.

Taking samples

For needle venepuncture, the needle should be held bevel uppermost and directed through the skin following the course of the vein. This should be completed in one movement. Once in the lumen of the vein, as determined by the presence of blood in the needle hub, the needle should be advanced parallel to the skin up to the hub, so that the body of the needle is in the lumen of the vein. The needle may be bent at an angle if required. (See Figure 10.2.)

A similar method can be used for butterfly needles and over-the-needle cannulae. For the latter, once the tip is within the lumen of the vessel, the flexible cannula can be advanced while retaining the stylet. This prevents the stylet from lacerating the vessel as it advances up the vein, and also the stylet blocks the lumen of the cannula so blood may only be withdrawn once the stylet is removed. Flexible cannulae are less traumatic to the tissues than needles, and are therefore less painful to insert and more suitable for long-term cannulation. (See Figure 10.3.)

Veins will collapse around the needle if attempts are made to withdraw the blood too quickly, so patience is required, particularly with small mammals. For these species, blood can be allowed to drip from a needle or flexible cannula placed in the vein. A syringe may be attached and gentle suction applied for larger animals. Evacuated

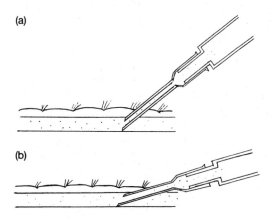

10.2 Method for superficial venepuncture. (a) The needle is inserted at an angle to the skin until the tip is in the lumen of the vein. (b) Once within the vein, the needle is advanced parallel to the skin until the hub of the needle reaches the skin.

tubes can be used if the vein diameter allows it. These are quick and easy to use, but can increase the likelihood of haemolysis as blood cells can become damaged by the rapid passage into the low pressure container.

After taking samples, gentle pressure should be applied to the site of venepuncture for 30–60 seconds to prevent haemorrhage.

Potential sequelae

Haemorrhage may occur from the punctured vein. If this occurs, pressure should be applied until the haemorrhage ceases. Dressings are available which if applied to a wound will accelerate haemostasis (e.g. Xenocol, Ichor).

Bruising may occur if the vein bleeds under the skin. Pressure should be applied as above, and the site rechecked after 30 minutes. If the bruised area continues to spread, advice should be sought from the named veterinary surgeon.

The occurrence of thrombosis or phlebitis following venepuncture indicates poor technique. The method should be reviewed and advice sought from the named veterinary surgeon.

Sample volumes

The volume of the sample taken is determined by the requirements of the experiment, and by the safe limit which can be withdrawn without causing distress to the animal. In general, as small a volume as possible should be taken. Not more than 10 per cent of the blood

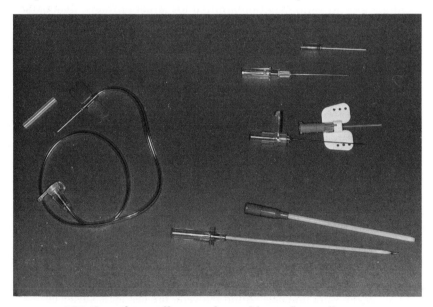

10.3 Over-the-needle cannulae and butterfly needle.

volume should be removed at one time, and less than 15 per cent of the blood volume should be removed in any 30 day period. Animals are considered to have 70 ml blood per kg bodyweight, but this varies with the species (see Table 10.1).

The volume and frequency of sampling will normally be stipulated on the project licence.

Methods of venepuncture
Rodents
For those rodents with tails, this is a good site for obtaining small quantities of blood. The lateral coccygeal or tail vein may be easily visualized, and the skin over the vein incised with a sterile lancet. Warming the tail first increases the blood flow to the site and makes sample collection easier. Blood may be collected into a capillary tube which can be obtained either plain or coated with anticoagulant, or allowed to drip into a container. Scalpel blades are not recommended as they can easily slip causing damage to the tail. Larger samples can be obtained from rats by placing a 23–26 gauge cannula into the tail vein and allowing blood to drip into a collection pot.

Amputation of the tail tip to obtain a sample of mixed arterial and venous blood can be carried out on rats, mice, and gerbils, but must be done under anaesthetic (BVA/FRAME/RSPCA/UFAW 1993). In mice, this does not appear to involve the removal of any vertebrae, which it does in rats. To follow the policy of the BVA/FRAME/RSPCA/UFAW working group on refinement, it should be performed only once, or a maximum of twice in mice.

Large samples can be obtained from rats and gerbils via jugular venepuncture. The vein is raised on one side of the neck by applying pressure at the thoracic inlet, and a needle placed through the skin and into the vein pointing towards the head.

For animals with no tails, such as guinea-pigs and hamsters, it is possible to obtain tiny samples by nicking the ear veins, but in practice it is usual to perform cardiac puncture, which must be done under general anaesthesia. There are many potentially harmful sequelae to this procedure, such as cardiac tamponade, and it should be done under terminal anaesthesia. The heart may be reached by placing the animal on its right side and piercing the left ventricle through the chest wall at the sixth intercostal space, one third of the way up, to obtain arterial blood, or by piercing the right ventricle with the animal on its left side for venous blood. Alternatively, the animal may be placed on its back, and the heart reached by passing the needle under the sternum and through the diaphragm. Jugular venepuncture although possible is extremely difficult in these animals. A technique has been described for cannulating the lateral saphenous vein in the hindlimb of guinea-pigs under general anaesthesia for repeated sampling (Nau and Schunk 1993).

Table 10.1
Blood sampling volumes

Species	Blood volume ml/kg	Total blood volume normal adult (ml)	Safe volume of single bleed[a] (ml)	Practical diagnostic volume (ml)	Bleed out volume (ml)
Small laboratory species					
Mouse	58.5	Male 1.5–2.4 Female 1.0–2.4	0.1–0.2	0.1	Male 0.8–1.4 Female 0.6–1.4
Rat	54–70	Male 29–33 Female 16–19	Male 2.9–3.3 Female 1.6–1.9	0.3	Male 13–15 Female 7.5–9
Guinea-pig	69–75	Male 59–84 Female 48–63	Male 6–8 Female 5–6	0.5	Male 29–42 Female 24–31
Gerbil	66–78	Male 4.5–7 Female 3.8–6	Male 0.4–0.7 Female 0.4–0.6	0.2	Male 2.2–3.5 Female 1.9–2.9
Hamster	78	Male 6.3–9.7 Female 7.1–11.2	Male 0.6–0.9 Female 0.7–1.1	0.3	Male 2.9–4.5 Female 3.3–5.2
Rabbit	57–65	58.5–585	5–50	1	31–310
Ferret	70	42–140	4–14	1	21–70

[a] A single bleed of 10 per cent of the blood volume averages 7 ml/kg.

Table 10.1 (cont.)

Species	Blood volume ml/kg	Total blood volume normal adult (ml)	Safe volume of single bleed[a] (ml)	Practical diagnostic volume (ml)
Laboratory primates				
Marmoset	70	21–24.5	2.1–2.4	0.5
Squirrel monkey	70	Male 39–77	3.9–7.7	0.5
		Female 24.5–52.5	2.4–5.2	
Rhesus	55–80	Male 420–770	Male 42–77	1–2
		Female 280–630	Female 28–63	
Cynomolgus	50–96	Male 280–560	Male 28–56	1–2
		Female 140–420	Female 14–42	
Baboon	62–65	Male 1430–1950	Male 143–195	1–5 plus
		Female 715–975	Female 71–97	
Larger domestic species				
Dog	70–110 [b]	900–1170 [c]	90–110	>1
Cat	47–65	140–200	14–20	>1
Pig				
Large white	56–69	13 200–15 000	1320–1500	>1
Yucatan	56–69	4200–4800	420–480	>1
Sheep	58–64	4060–4480	400–450	>1
Goat	57–90	3990–6300	400–630	>1
Cattle	60	27 000–36 000 [d]	2700–3600	>1
Horse	75	33 750–45 000 [d]	3375–4500	>1

[a] Single bleed of 10 per cent of blood volume. [b] Much breed variation. [c] Beagle. [d] Assumes adult weight 450–600 kg.

Collection of blood from the orbital venous sinus can have severe consequences for the animal, and although it has been used for bleeding rats and tail-less animals, it is not recommended for sampling with recovery.

Rabbit

Blood can be collected relatively easily from the marginal ear vein using an over-the-needle cannula or butterfly needle. A peripheral vasodilator may be applied to the skin over the vein (e.g. Vasolate, IMS), 5–10 minutes before blood collection. Once the vein is engorged, the cannula is inserted and blood can be collected by allowing it to drip into a pot. After collection, the vasodilator is wiped off and gentle pressure applied until the bleeding ceases. Haemostatic dressings can be applied as discussed earlier in this chapter.

Bleeding from the central ear artery is possible, but can result in the formation of large haematomata which can cause damage to the ear or even necrosis.

Ferret

For tiny quantities of blood, e.g. for an Aleutian disease test, a toenail can be clipped and a drop of blood collected into a capillary tube.

Larger quantities can be collected from the jugular vein. The fur on the neck needs to be well clipped, and the vein raised by placing a thumb over the jugular groove in the thoracic inlet. Blood is collected by inserting a needle up the vein towards the head, or down towards the thoracic inlet. Collection is facilitated by bending the needle to an angle of 30° prior to penetrating the skin. Pressure on the vein in

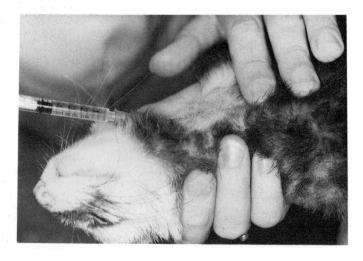

10.4 Jugular venepuncture in the ferret.

the thoracic inlet is maintained until the blood has been collected. (See Figure 10.4.)

Primates
The best method of blood withdrawal in primates is to use the femoral vein, in the groin. This can be used for Old World and New World monkeys. The needle is inserted in the femoral triangle, slightly medial to the femoral pulse, pointing towards the head. (See Figure 10.5.)

For larger Old World monkeys, the cephalic vein on the top of the foreleg below the elbow can be used, as for cats and dogs. The jugular vein can be used as an alternative route.

Marmosets can also be bled from the coccygeal vein.

Dog and cat
These can be bled from the jugular vein. A handler places their right arm over the body of the animal to hold the forelegs. The elbow is used to hold the body of the animal to the body of the handler. The left hand is placed under the chin to raise the head. The person collecting the blood raises the vein by placing a thumb in the jugular groove at the thoracic inlet. (See Figure 10.6).

The cephalic vein can also be used (see Chapter 9 on handling).

Ruminants and Horses
For sheep, goats, cattle, and horses the jugular vein is used. The vein can be visualized by clipping hair or wool from over the jugular

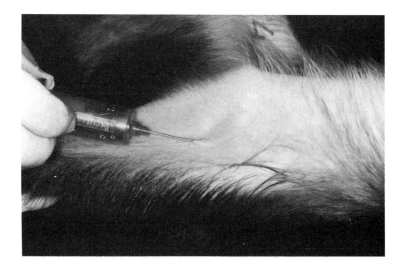

10.5 Femoral venepuncture.

groove. The vein is raised by applying firm pressure to the base of the jugular groove. The needle is advanced through the skin up towards the head. (See Figure 10.7a and b.)

Cattle can also be bled from the ventral coccygeal vein.

Pig
These are probably the most difficult animals to bleed. For small volumes in large pigs, an ear vein can be used. For larger volumes, the anterior vena cava is used. A long needle is inserted in the thoracic inlet, and angled slightly upwards and medially to enter the anterior vena cava.

Birds
For chickens, the brachial vein is used. It can be visualized as it crosses the elbow by plucking the feathers over the medial surface of the wing. Samples should be removed slowly to prevent the vein collapsing. Pressure is generally not applied after venepuncture in birds as this can promote haematoma formation.

The right jugular vein can also be used. It is found between the feathers on the dorsolateral surface of the neck. It can be raised by applying pressure at the base of the neck. Haematomata rarely form after jugular venepuncture.

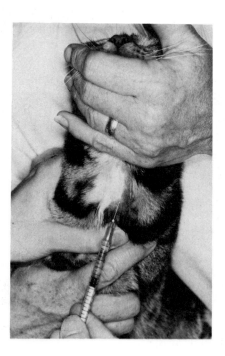

10.6 Jugular venepuncture in the cat.

ADMINISTRATION OF SUBSTANCES
Introduction to pharmacokinetics
Administration and absorption

Substances may be administered to laboratory animals by mouth, by intravenous, intramuscular, intradermal, intraperitoneal or subcutaneous injection, per rectum, or by injection directly into other body parts such as joints or parts of the gastrointestinal system. The substances may have effects locally or at distant sites after absorption into the bloodstream. The rate at which administered compounds are absorbed into the bloodstream depends on the site of administration, the nature of the compound, and the manner in which it is presented.

Compounds which are given intravenously reach a high blood concentration immediately, which then tails off as the compound is eliminated, by the liver and kidneys or other routes, or redistributed, e.g. by absorption into fat. Compounds given by other routes are absorbed at rates depending on the blood flow to the site, and the solubility of the compound in the tissue fluids. Muscles have a good blood supply, so substances administered intramuscularly are absorbed more quickly than those given subcutaneously, as the subcutis has a poorer blood supply. Compounds which are highly

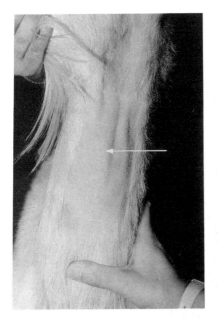

10.7 (a) Visualizing the jugular vein in the goat (arrow). (b) Jugular venepuncture in the goat.

soluble in the tissue fluids will be absorbed quickly. Some injectable preparations are designed to have a long duration of action, and the active compound in these is mixed with a carrier of low solubility to slow down absorption into the bloodstream. With orally administered compounds, absorption tends to be slower, and takes place over a longer period. (See Figure 10.8.)

Compounds which are given orally may be absorbed at various points in the gastrointestinal tract. Substances will only be absorbed across the gut wall if they are lipid-soluble. Some compounds are designed to be absorbed locally in the large bowel, and are insoluble unless they are activated by enzymes during passage through the stomach and small intestines. The absorption of other compounds may be affected by pH. Compounds which are ionized are not lipid soluble and will only be absorbed if a specific carrier exists to transport them across the gut wall into the blood. Basic or alkaline compounds are likely to be fully ionized in the acid environment of the stomach, and will be poorly absorbed. However, once in the small intestine, the higher pH will reduce the level of ionization and increase the absorption. The reverse is true for acidic compounds.

For solids, the particle size affects the rate of absorption, because the surface area relative to volume increases as particle size decreases, presenting more of the compound for solubilization and absorption (see Figure 10.9).

Another factor which complicates oral administration is the **first**

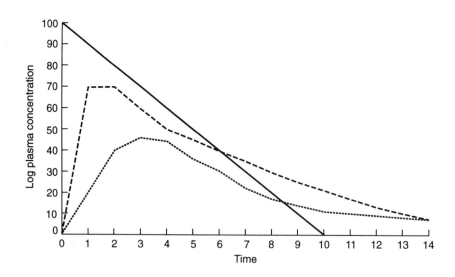

10.8 Decline of plasma concentration after intravenous (i.v., ——), intramuscular (i.m., ------), and oral administration (·······) of a typical compound.

pass effect. All substances absorbed in the stomach or intestine, excluding the mouth or rectum, have to pass through the liver before reaching the systemic circulation. This may result in the metabolism of some or all of the compound to active or inactive products, reducing the amount of the original compound which reaches the circulating blood volume. Many compounds induce the liver to produce enzymes which metabolize them, thereby increasing the first pass effect with repeated dosing. Alternatively, there may be some **enterohepatic recycling**, in which the compound is conjugated and secreted into bile, and therefore passed back into the intestine without reaching the systemic circulation.

These factors affect the **bioavailability** of substances administered orally, which is essentially the difference between the amount of an oral dose which reaches the bloodstream compared with an equivalent intravenous dose. However, even if the bioavailability is high, and most of an oral dose is absorbed into the bloodstream, it does not mean that the blood concentration reaches the peak seen immediately after an intravenous dose. Many compounds are absorbed very slowly after oral administration, and never reach high levels in the blood even if the bioavailability is high.

Distribution
The sites of distribution of compounds in the body depend on the same factors as absorption, i.e. the blood flow to the site and the

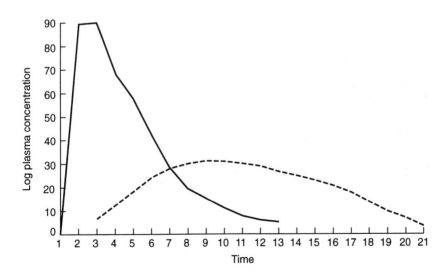

10.9 Plasma concentration after oral administration with compound administered in small (———) or large (-------) particles. (The bioavailability equals the area under the curve.)

solubility of the compound in the tissue fluids. There are some special cases, however, e.g. compounds will only enter the CNS if they can cross the blood–brain barrier. This is formed by cells around the capillaries which prevent substances from entering the tissues except at the choroid plexus, and even there only certain compounds will cross.

Lipid soluble compounds will accumulate in fatty tissues, even if the blood supply is poor. This phenomenon can be used to ensure rapid elimination of lipid soluble drugs from the CNS by redistribution to fat (see thiopentone in Chapter 12). Drugs will also tend to accumulate at sites of metabolism or excretion, such as the liver and kidneys. In pregnant animals, the fetus is separated from the mother by a semi-permeable membrane so only certain compounds will enter the fetus. Compounds will generally be secreted into milk if they can cross the lipid membrane in the mammary gland. Milk is slightly more acidic than blood, so basic compounds tend to accumulate in milk as having crossed the lipid barrier they ionize and cannot then return to the blood.

These factors which affect the absorption and distribution of administered compounds define the pharmacokinetics of the compound, and should be studied prior to administration to ensure that the compound is being given in a logical manner.

Administration volumes

The volume of any substance given must be as small as is practicable for the procedure, and will be limited ultimately by the size of the animal. If fluids are to be administered by infusion, the flow rate should be as low as possible, and the infusion given over as short a time as possible. If too much fluid is given too rapidly, the circulation may become overloaded, causing pulmonary oedema. If the administration is slow, excess fluid can be cleared by the kidneys. Care should be taken in normal animals not to exceed the maintenance requirement for fluid, which is approximately 40 ml/kg per 24 hours. (See Table 10.2.)

For large animals, (sheep, goats, and pigs), much larger volumes can be given safely. The guidelines for dogs can be followed, increased proportionally with the increased bodyweight, up to a maximum of 7 ml per site for intramuscular injections.

Injection techniques

Injections should be performed using aseptic techniques, as for blood sampling. It is important to use equipment appropriate for the species. For example, injections in small rodents should be given with 25 or 27 gauge needles. For rabbits, guinea-pigs, and cats, 23 or 25 gauge needles are best, and for dogs 21 gauge needles are adequate. For farm animals, needles larger than 21 gauge may

Table 10.2

Administration volumes for laboratory species

Species	Intravenous or arterial (ml)	Intraperitoneal (ml)	Intramuscular (ml/site)	Subcutaneous (ml/site)	Oral (ml/kg)	Intradermal (μl/site)
Mouse	0.2	2–3	0.05	0.5 [a]	20	100
Rat	1	5–10	0.1	1–2 [a]	10	100
Guinea pig	0.5	10–15	0.3	1–2 [b]	5	100
Hamster	0.3	3–4	0.1	0.5–1 [b]	20	100
Rabbit	1–10	50–100	0.5–1	1–5 [b]	5	100
Dog	10–15 [c]	200–500	2–3	15–30 [b]	3	100
Cat	2–5	50–100	1	5–15 [a]	3	–
Ferret	1–3	5–10	0.5–1	1–2 [b]	5	–
Fowl	2–3	10–15	1–2	–	–	–
Rhesus (5 kg)	5–10	50–100	1–2	5–10 [a]	10	100

[a] Maximum of 2–4 sites. [b] Maximum of 4 sites. [c] Beagle.

Table 10.3
Suggested hypodermic needle size (gauges) for laboratory animals

Species	Intraperitoneal	Intramuscular	Intravenous	Subcutaneous
Mouse	27	27	27–28	25
Rat	23–25	25	25–27	25
Guinea-pig	21–25	25	25–27	23–25
Gerbil	27	27	27–28	25
Hamster	23–25	25	25–27	25
Rabbit	21–23	23–25	23–25	21–25
Ferret	21–23	23–25	21–25	21–23
Dog	21–23	21–23	21–23	21–23
Cat	21–23	23	21–25	21–23
Rhesus	21–23	23–25	21–25	21–25
Sheep	19–21	21	19–21	19–21

be used. (See Table 10.3.) The viscosity of the substance also affects the size of needle used. Thick, viscous liquids may not pass through narrow gauge needles. Intradermal injections are performed with 25–27 gauge needles.

To minimize the distress caused to animals during administration of substances, they must be carefully and expertly handled, and given a sedative or general anaesthetic if required.

Subcutaneous injections
For most species, subcutaneous injections can be given into the scruff of the neck (See Figure 10.10). A fold of skin is lifted using the thumb and first two fingers of one hand, and the needle is passed through the skin at the base of the fold parallel to the body, to avoid penetrating deeper tissues. Subcutaneous injections are rarely painful.

In rabbits, subcutaneous injections can also be given over the flank, provided that care is taken with adjuvants, because if there is an adjuvant reaction in the skin over the respiratory muscles this can cause pain on breathing. In sheep and goats, fluid can be given under the skin over the ribs. For pigs, small volumes can be injected into the skin behind the ear, or into the fold between the leg and the abdomen. Pigs have much subcutaneous fat, and injections given elsewhere are likely to enter the fat, where absorption will be particularly slow due to the poor blood supply.

Intramuscular injections
Intramuscular injections are frequently painful, due to the distension of muscle fibres which occurs, and therefore good technique and restraint are required. They are usually given into the muscles of

the thigh. Large volumes or potentially irritant compounds should be injected into the quadriceps group on the front of the thigh. The muscle can be immobilized with one hand while injecting with the other. Injections can be given into the caudal thigh muscles, but as the sciatic nerve runs through these muscles, irritant compounds should not be given here or damage may be caused to the nerve. In rodents, the quadriceps feels like a small peanut on the front of the thigh, and can be immobilized with the thumb and forefinger of one hand whilst injecting with the other. In dogs and cats, injections can be given with care into the muscles on each side of the spine, and in large animals the gluteal muscles are used. In adult pigs, injections are given into the neck muscles, but a long needle is required to penetrate the fat layer. Piglets can be injected by lifting them by one hindleg and injecting into the caudal thigh muscle on that side. Fowl are given intramuscular injections into the pectoral muscles. After the injection, the site should be massaged to disperse the dose. (See Figure 10.11.)

Intravenous injection

Intravenous injections may be given into the cephalic veins of dogs, cats, primates, and ferrets. In rats and mice, the lateral tail vein is used. Rats can also be given injections into the dorsal metatarsal vein. The hind limb is held in extension, and the vein raised by occluding it at the stifle joint. The jugular vein can be used in dogs, cats, and hamsters, and is the method of choice in ruminants and horses. The ear veins are used in guinea-pigs (Figure 10.12), rabbits, and pigs. Fowl can be injected via the brachial vein.

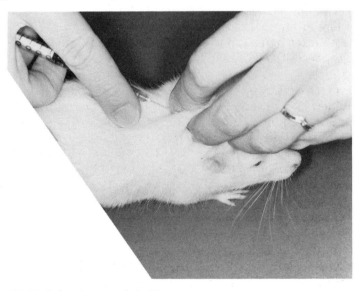

10.10 Subcutaneous injection.

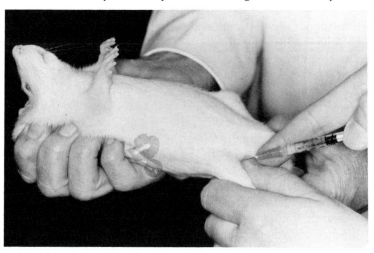

10.11 Intramuscular injection.

Intradermal injections

Intradermal injections for most species can be given in the same area as subcutaneous injections. For tuberculin testing in primates, the skin of the upper eyelid is often used.

Intraperitoneal injections

Intraperitoneal injections in rodents are given in the lower left or right quadrant of the abdomen as there are no vital organs in this area. The quadrants are demarcated by the midline and a line

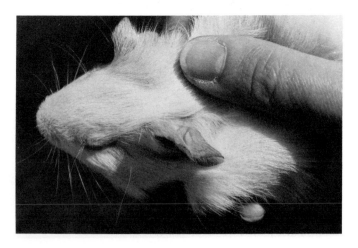

10.12 Guinea-pig ear vein.

perpendicular to it passing through the umbilicus. The animal should be held either by an assistant or in one hand on its back, and the needle angled at 45° to the skin. No resistance should be encountered to the passage of the needle. (See Figure 10.13.)

Oral administration

Substances may be given orally by inclusion in the diet or drinking water. These methods have the disadvantages that it is not possible to be sure that the animal has had the entire dose, and in some species it will lead to the animal refusing to eat and drink. With *ad libitum* feeding, or if there is an increase in metabolic rate, the animal may overeat and thus ingest an overdose of the drug. In mice particularly, adding drugs to the water tends to lead to dehydration, because the mouse avoids drinking it, and this can lead to a rapid deterioration in the condition of the mouse, especially if the compounds have been given for therapeutic reasons. If the watering system is automated, it is not possible to give compounds in this way. To overcome these problems, gastric intubation or gavage may be employed (Figure 10.14). Flexible catheters or stainless steel needles with rounded tips are used. The animal is restrained with its neck extended, and the needle or catheter passed gently down

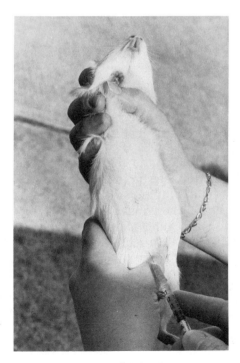

10.13 Intraperitoneal injection.

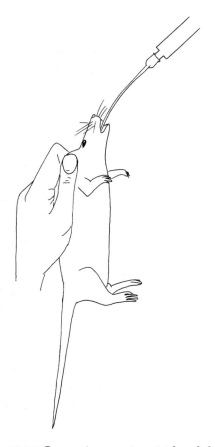

10.14 Gavage in rat using rigid oral dosing needle.

the oesophagus. Care must be taken not to damage the oesophagus, or to put the needle into the trachea. The needle or catheter can usually be observed passing down the oesophagus on the left side of the neck. Damage to the catheter from chewing can be avoided by using an oral speculum, or by using a flexible nasogastric or pharyngostomy tube instead. The animal may or may not need to be starved prior to administering the compound, depending on the nature of the compound and the particular project.

For dogs and cats, tablets may be administered orally in the conscious animal by placing one hand over the top of the head, placing the thumb at the commissure of the lips on one side and the fingers at the other, and tilting the head back. This will cause the mouth to open slightly. The tablet can be held between the thumb and forefinger of the other hand, and the middle finger used to open the mouth. The tablet can then be placed onto the tongue as far back as possible, to stimulate the swallowing reflex.

INDWELLING CANNULAE, ARTERIAL LOOPS, AND FISTULAE

Placement and maintenance of indwelling cannulae

It is possible to insert cannulae into the veins or arteries of almost all species, except mice. Butterfly needles may be used, but flexible plastic cannulae are better as butterfly needles may lacerate the vein if there is any movement.

Cannulae may be implanted surgically, or passed into a superficial vein by puncture of the skin, as for needle venepuncture. To prevent clotting in the cannula, it should be flushed with heparinized saline or another anticoagulant after placement, and between sampling or administering compounds. A concentration of 10 to 15 units heparin per ml saline is adequate. If there is a three-way tap or multiple sampling port on the end, this is made much easier as the anticoagulant delivery and sampling/administrating systems may be attached at the same time.

Maintenance of cannulae for repeated administration is easier than for repeated withdrawal, as blood clots on the end of the cannula tend to allow administration but act as valves preventing withdrawal.

Withdrawal of blood

When withdrawing blood from a cannula, the first sample taken each time should be discarded, as it will contain anticoagulant and be diluted. After sampling, the cannula should be flushed with a calculated amount of anticoagulant: enough to fill the cannula, but not to enter the bloodstream.

Long-term cannulation

If cannulae are maintained for longer than 24–48 hours, blockages due to thrombi can become a serious problem. Long-term cannulae are usually implanted surgically, which must be done under aseptic conditions and requires considerable skill. In young animals, room must be left so the cannula is not pulled out as the animal grows, and in all animals movement can be a problem. The strategic positioning of lengths of flexible tubing attached to the cannula can reduce the likelihood of it being dislodged by movement.

Cannulae can be placed in a major vessel, or into a minor one with the tip passing into a major vessel. Cannulae placed in the jugular vein should not enter the right atrium, or arrhythmias may be caused.

Cannulae should end in sealed ports, which should be capable of either being pierced several times, or removed readily to allow administration or withdrawal under aseptic conditions. The cannula may be tunnelled through the subcutis to exit the skin at a distant site, e.g. over the back, to reduce interference from the animal and the risk of ascending infection. Such cannulae should not kink or pull out when the animal moves. Exit sites are potential sources of infection and irritation, and must be inspected at least once daily.

Cannulae can be kept patent for more than three or four weeks in small animals, and longer in large animals. The period can be increased if the animal is restrained, but this results in distress to the animal and is best avoided.

Cannulae need to be of a suitable inert material which will not cause any irritation, and be strong enough to withstand kinking and occlusion due to movement or tightening of ligatures. Polypropylene cannulae are strong, but silicone rubber causes less tissue reaction. The two can be combined by placing a silicone rubber cannula over a polypropylene one, or by coating a polypropylene cannula with silicone rubber.

Removal of cannulae
After removal, haemostasis can be achieved by applying pressure to the vessel for several minutes, or by ligating the vessel with a suitable suture material (see Chapter 15).

Arterial sampling and arterial loops
Arterial sampling allows very large samples to be taken rapidly. Often, the carotid artery is used, but the central ear artery can be used in rabbits, or the femoral artery in many other species.

Needle puncture. Good restraint is essential when taking samples with a needle as accidental movement can result in laceration of the artery and profuse haemorrhage. Once the needle is in the artery, the high blood pressure will normally force the plunger in the syringe back. Firm pressure should be applied to the site for 2–5 minutes afterwards to prevent haematoma formation.

Arterial cannulae. When introducing arterial cannulae, the artery should be clamped with atraumatic vascular clamps, as otherwise there will be haemorrhage as soon as the artery is punctured. Cannulae should be placed in the direction of the flow of blood.

There is a greater potential for thrombi·to cause problems in arterial cannulae than venous ones, as thrombi may embolize and cause infarctions in distant sites such as the renal artery (see Figure 10.15).

Arterial loops
This method involves bringing an artery into a subcutaneous position for easy sampling. This has to be done surgically, and is

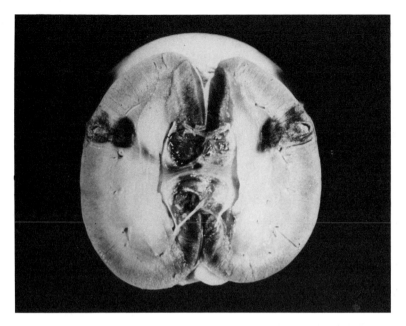

10.15 Renal infarction due to thrombus.

technically difficult. This must first be practised on cadavers and then performed under very close supervision.

Maintenance of fistulae
A fistula is an artificial passage created between two hollow organs, or between a hollow organ and the outside. For example, ruminal fistulae may be created to study the physiology of the rumen in sheep, goats, and cattle, or for the treatment of bloat. Fistulae are created surgically under general anaesthetic. An incision is made in the skin overlying the organ, and the organ is brought up to the surface. An incision is made in the wall of the organ, and the skin edges are sutured to the cut edges of the organ. Aftercare involves checking and cleaning daily to prevent early closure, or damage to the skin around the fistula from exudation of serum or organ contents. At the end of the experiment, the fistula may be closed surgically, or allowed to heal over by second intention healing.

MISCELLANEOUS METHODS OF ADMINISTRATION AND SAMPLING
Administration
Footpad/tail base injection
Substances such as Freunds Complete Adjuvant (FCA) may be injected into the foot pad or tail base of rodents to induce localized

or generalized arthritis respectively. Animals with arthritis **must** be given soft bedding, and may be provided with food and water in a gel placed adjacent to the bed to avoid the necessity for them to walk around. Social animals should be kept in groups but at a low stocking density so they do not have to climb over each other to reach food or water.

Nebulization/administration into the respiratory tract
Some liquids may be nebulized into tiny particles for administration into the respiratory tract. This may be necessary for some drugs and vaccines. The compound is given in a spray which is inhaled, or in some cases absorbed through the skin. The spray may be administered to the entire body, or just to the nose.

Inoculation into the eyes
Compounds for topical ocular administration must be non-irritant, or there may be discomfort, lacrimation, and corneal damage, with scarring in the long term.

Intra articular injection
Injections into joints must be carried out under aseptic conditions to avoid causing septic arthritis. Compounds may be administered by this route for experimental or therapeutic reasons, or samples of joint fluid may be taken for analysis. It is essential to be certain of the anatomy of the joint and its synovial capsule before penetrating the skin, so as to be certain of entering the correct part of the joint. The procedure should be practised on cadavers first.

Per rectum
Enemas may be given to empty the rectum before bowel surgery or radiography, or fluids and some drugs may be given into the rectum or distal colon. Substances may be given via a flexible rubber or polythene tube, or a rigid sigmoidoscope. Whichever method is used, it is essential that the instrument is well lubricated and inserted very gently.

Sampling
Urine
Urine may be collected as it is voided naturally in all mammals, or by using various techniques. A metabolism cage, in which all the animal's waste products can be collected, can be used in all species, and in rodents this is the only reliable method for obtaining a sample. For larger species, urine may be expressed manually (e.g. cats), collected by using a urethral catheter if the urethra is wide enough (e.g. rabbits and larger species), or withdrawn by cystocentesis under aseptic conditions. With the latter method, the skin over the lower

abdomen is clipped and cleaned as for surgery, the full bladder is immobilized with one hand and a sterile needle of a suitable gauge (see Table 10.3) inserted through the body wall into the bladder.

Urinary catheters may be maintained *in situ* for prolonged periods, either after surgical implantation (in which the catheter may be channelled to exit the skin at a distant site as for venous cannulae), or passing into the urethra. It is important with indwelling urinary catheters to avoid urine scald, by collecting the urine into a bag, or by ensuring meticulous cleanliness around the end of the catheter and protecting the skin with petroleum jelly or a similar barrier. It is also important to prevent infection from tracking up the catheter into the urinary tract. This can be achieved by ensuring good hygiene and by prophylactic use of appropriate antibiotics. The urine can be monitored bacteriologically to ensure that there is no infection.

Faeces

Faeces may be collected after being voided naturally, by using an enema, or by using a rectal swab. A metabolism cage can be used, and in rodents again this is the most reliable method. Enemas may be given to collect large quantities of faeces, or swabs may be used for small samples, e.g. for bacteriology. It is essential that anything passed into the rectum is inserted gently.

In rodents and rabbits, which are coprophagic, hard faeces can be collected as above, but to collect soft faeces, which are normally eaten straight from the anus, restraining devices must be used.

Semen

Semen can be collected from the larger animals in a number of ways. In cattle and horses, a teaser female can be used to excite the male, and an artificial vagina used to collect the semen. In rams, electro-ejaculation is usually used. A probe is inserted gently into the rectum, and an electric current stimulates the nerves to the genitals resulting in ejaculation. In dogs, digital manipulation can be used, with or without the presence of a bitch in season.

Milk

The mammary glands of ruminants and horses are served by teats, whereas those of the smaller animals have nipples. Teats are larger, with one or two orifices from which milk can be expressed, and the udders of these species are large enough to permit the taking of large samples. At peak lactation, up to 17 litres of milk may be withdrawn twice daily from a Friesian cow. Nipples have several orifices from which milk flows, and small samples can be collected from these.

Milk can be expressed by grasping the nipple or teat at its base and gently pulling towards the tip. In newly parturient females, injections of oxytocin may increase the release of milk.

POLYCLONAL ANTIBODY PRODUCTION IN LABORATORY ANIMALS
Polyclonal antibody production using adjuvants
Production of antibody may be achieved by injection of the required antigen into an animal, in combination with a suitable adjuvant. Adjuvants stimulate the immune system to produce the appropriate class of antibody, by prolonging exposure of the antigen to the immune system, and promoting phagocytosis and antigen processing.

A commonly used adjuvant is Freunds Complete Adjuvant (FCA), in which paraffin oil is used to produce the depot effect, and killed *Mycobacteria* are used to stimulate macrophage responses and antigen processing. Unfortunately, this adjuvant often produces severe inflammatory reactions, such as tissue swelling, necrosis, abscessation, and sloughing. Extreme care should be taken in preparing and using FCA in order to reduce the occurrence of such reactions. If used correctly, FCA can produce an exceptional immune response with a minimum of tissue damage.

Several other adjuvants are available which produce less severe reactions, e.g. Hunters Titer-Max. For further information see Further Reading.

Protocol
The initial immunization can be performed using the antigen adjuvanted with FCA, but thereafter booster injections should contain Freunds Incomplete Adjuvant (FIA). This differs from FCA in that the mycobacterial component is absent.

The two major factors determining the severity of the reaction are injection site, and injection volume.

1. *Injection site.* FCA and FIA may be given intradermally, subcutaneously, or intramuscularly. In mice, adjuvants may be given by the intraperitoneal route, but the resulting ascites is painful and strict controls on numbers used and adverse reactions must be observed.

Foot pad immunization in rodents should only be performed if study of the draining lymph node in isolation is required. If this is necessary, only one foot should be immunized per animal, and soft bedding should be provided. Injection of FCA into the tail base of rodents can be used to induce polyarthritis.

With the intradermal (i.d.) route, the depot effect produced by the adjuvant is prolonged. The antigen is protected from degradation and is undispersed, remaining exposed to the immune system for longer. Unfortunately, large non-painful granulomas with ulceration of the skin develop, which may become painful abscesses if traumatized by the animal or if the injection is not sterile. The severity of the reaction to i.d. injections varies with the injection volume. However, ulceration of the skin occurs routinely with all dosages.

Subcutaneous injections are easier and produce fewer granulomas. However, with subcutaneous administration, the adjuvant tends to track ventrally, producing fistulous tracts and firm linear lesions distant from the injection site. Care must be taken not to give the injection intradermally or intramuscularly inadvertently (see Figure 10.10).

With intramuscular injections, although any lesions developing are not readily detectable, they are particularly painful when the muscles are used, e.g. for respiration or locomotion. Therefore subcutaneous administration is preferable.

FCA and FIA should not be given by the intravenous route, since this results in generalized abscessation and death.

Strict adherence to sterile technique when giving the injection is essential, to reduce the occurrence of bacterial infections. The skin should be clipped of fur and swabbed with alcohol before the injection, and a new sterile syringe and needle must be used. The normal adjuvant-induced granulomas are composed mainly of macrophages and lymphocytes, and are non-painful. The introduction of bacteria causes an infiltration of polymorphonuclear leucocytes, which tends to be painful. This pain predisposes to self-mutilation by the animal.

2. *Injection volumes.* The initial immunization using FCA should not exceed the recommendations in Table 10.4.

Booster injections using FIA should also follow these recommendations. They should be given at intervals such that complete recovery from one injection is allowed before the next is given. A period of 30–45 days is adequate for this.

The final dose injected should be composed of not more than 50 per cent FCA or FIA, mixed with antigen. To facilitate the preparation and injection of aqueous antigen mixtures, the emulsion can be prepared with equal parts of FCA, antigen, and mineral oil.

Polyclonal antibody production using perforated subcutaneous chambers

With this method, a plastic practice golf ball is implanted surgically under the skin of the lateral thorax of rabbits. The centre of the ball fills up with transudate, which can be withdrawn simply by perforating the skin through into the centre of the chamber using a hypodermic needle and syringe, under aseptic conditions (Clemons *et al* 1992).

After implantation, the ball is left for 4 weeks in order for it to become encapsulated. Thereafter, the chamber can be inoculated percutaneously using antigen **without** adjuvant, and boosted as

Table 10.4
Injection volumes for Freunds Complete (FCA) and Incomplete
Adjuvant (F1A)

Route	Species	Volume of single dose (ml)	Volume of maximum total dose (ml)*
Intradermal	Rabbit	0.05–0.1	0.4
Intramuscular	Rabbit	0.1–0.25	0.5
Subcutaneous	Guinea-pig	<0.25	<1
	Mouse	<0.1	<0.2
	Rabbit	0.1–0.25	1.0
	Rat	<0.2	<0.4
	Sheep and goat	<0.25	<1

* In rabbit, guinea-pig, sheep, and goat a maximum of four sites should be used, and in rats and mice a maximum of two sites should be used.

required. Antibody can be harvested using a needle and syringe. A second filter-hubbed needle placed through the skin into the chamber prevents any damage due to negative pressure when fluid is withdrawn.

The volume of antibody-rich fluid which can be withdrawn from the chamber weekly varies from 10 to 20 ml. Antibody levels decline faster than those in traditionally immunized animals, but the problems associated with repeated blood sampling are avoided, and the fluid can be harvested rapidly with little distress to the animal.

MONOCLONAL ANTIBODY PRODUCTION

Monoclonal antibodies can be manufactured by the fusion of antibody-producing cells and myeloma cells, to produce a set of cells known as **hybridoma** cells. These hybridoma cells can be cloned in large quantities, and will secrete antibody with a particular specificity and affinity (monoclonal antibody) for a single antigen. Standard methods for antibody production result in the formation of antibodies with variable specificities and affinities for a single antigen (polyclonal antibody).

The injection of hybridoma cells into the peritoneal cavities of mice results in the formation of higher titres of monoclonal antibody in the ascites fluid than are obtained from the serum of animals immunized by other means, or from cell culture. This makes antibody production using hybridoma cells in mice very popular. However, this method is not recommended unless fewer than 20 mice are to be used and the procedure is performed only once. *In vitro* methods are expected to be used where possible for the production of monoclonal antibodies.

To increase the volume of ascites fluid produced, the abdomen may first be 'primed' with pristane or Freunds Incomplete Adjuvant (FIA). These cause irritation to the peritoneal cavity resulting in fluid production. The irritation causes the mouse distress, leading to a hunched appearance, weight loss, and inactivity. To minimize these effects, the priming injection of pristane injected into the abdomen should be less than 0.2 ml in volume. For FIA, it should be less than 0.1 ml. With pristane, hybridoma cells are injected a few days after priming, and antibodies can be harvested 2 to 3 weeks after the initial pristane dose. With FIA, hybridoma cells can be inoculated the next day, and antibodies harvested within 14 days. Antibodies can be harvested **once** with recovery, using a suitably sized needle, under general anaesthesia, but ideally the procedure should be terminal. The mouse should be watched closely for signs of hypovolaemic shock after withdrawal of the ascites fluid, as the volume taken can exceed the blood volume of the mouse.

The animals should be monitored daily following the priming injection for signs of distress, and euthanased upon reaching a defined humane end point (see distress scoring sheet, Figure 11.1). The volume of ascitic fluid should not be allowed to exceed 20 per cent of the normal bodyweight.

11 Recognition of pain and stress in laboratory animals

To satisfy the provisions of the Animals (Scientific Procedures) Act 1986 an assessment of suffering is required for consideration of the cost benefit analysis and to assess the level of severity in the project licence application [see Chapter 2, summary of the main provisions of the Animals (Scientific Procedures) Act 1986].

It is the responsibility of the personal licence holder to look after the health and welfare of his/her animals. There will be specific conditions attached to the project licence under which the work is done which will limit the amount of pain and discomfort any animal may experience. **The use of the *humane end point* in the refinement of the experiment will help to reduce any suffering experienced by the animals.** It is inevitable that some laboratory animals will experience discomfort as a result of the scientific procedures carried out on them, but pain is an unnecessary accompaniment to the majority of scientific procedures.

Every effort must be made to identify the causes and to control the adverse effects of scientific procedures upon animals. It is important to realize that in addition to humane considerations the presence of pain can produce a range of undesirable physiological changes which may alter the rate of recovery from surgical procedures, and these changes may affect the results obtained.

The first stage in assessing an animal's well-being is to become familiar with its normal appearance and behaviour. When doing 'hands on' training in the animal house take time to observe the animals that you will be using in order to become familiar with their normal behavioural repertoire in the laboratory environment (as compared to pet animals, for example, hamsters behave differently in a 'Rotastak' environment compared to a laboratory 'shoebox').

WHAT IS STRESS?

Physiological stress   occurs within normal physiological limits. The animal uses minimal effort to respond and is unconscious of this effort.

Overstress   this requires significantly more effort but the animal is still unconscious of it. It can be detrimental to biological processes such as growth.

Distress   considerable effort has to be put into the response, of which the animal is aware. The animal can be considered to be suffering.

Pain   is an unpleasant sensory and emotional experience associated with actual or potential tissue damage. It can be divided into acute or chronic depending on the time scale over which it occurs.

Three different types of response to pain will be:

1. To modify conscious behaviour to avoid repetition of the painful situation (requires high level CNS function).
2. Automatic responses to protect the animal or a part of it, for example, withdrawal reflex, freeze or flight responses.
3. Response to convey the experience to others in the group thus ensuring survival of some of the population. This may occur as vocalization or by the release of pheromones, and may cause stress to other animals nearby, especially if they are unable to react as their normal behavioural repertoire would demand.

HOW IS PAIN ASSESSED?

An animal's response to pain will be modified by a number of factors. Consider:

1. The *individual details* of the animal such as its species, age, and origin. These factors will all affect its response to painful stimuli.
2. The *history* of the animal and the establishment. Take into account previous problems encountered, the course of current problem, the environment in which the animal is kept, the procedures that have been carried out, and any known current disease problems.
3. *Clinical examination* of the animal to assess its current

condition, and to find out if there is intercurrent disease (if so contact the named veterinary surgeon for advice, see Chapter 7).

Take note of
– physiological signs (such as heart rate, respiration rate, body temperature, muscle tone),
– biochemical signs (such as ACTH levels, endorphins)
– feeding behaviour (for example quantity, pattern of feeding).

4. *Mental status.* Take notice if the animal appears dull, depressed, aggressive, or hyperexcitable, especially if such traits are at variance with its usual behaviour. The technician in charge of the animal will often be the best person to observe any changes in its temperament or its reluctance or otherwise to accept handling.

5. The *activity* of the animal may range from total inactivity to maniacal hyperactivity. Notice if there are any changes in gait or posture or facial expression.

6. *Vocalization* will depend on the species and there are a wide variety of different noises produced by each species. The sound emitted may be outside the human auditory range and therefore go unnoticed but may be causing distress to others of the same species.

7. *Response to analgesics.* If a dose of an analgesic drug is administered and the animal's condition and demeanour improves, then this can be a useful diagnostic aid in assessing whether pain was present.

These are very general descriptions on how to assess pain qualitatively, and it is necessary to consider how to quantify pain in order to be able to judge whether it has been alleviated and to ascertain whether the degree of pain is within acceptable limits for the severity banding of the project.

HOW TO QUANTIFY PAIN (see Figure 11.1: Distress scoring sheet)

A range of clinical signs are assessed, and given a score, and the overall score indicates the likelihood of whether the animal is suffering. For a simple pain scoring system, monitor:

- Appearance
- Food and water intake
- Clinical signs
- Natural behaviour
- Provoked behaviour

ANIMAL IDENTIFICATION	Score	Date/Time	Date/Time	Date/Time	Date/Time	Date/Time	Date/Time	Date/Time	Date/Time
APPEARANCE									
Normal	0								
General lack of grooming	1								
Coat staring, ocular, and nasal discharges	2								
Piloerection, hunched up	3								
FOOD AND WATER INTAKE									
Normal	0								
Uncertain: body weight ↓ <5%	1								
↓ intake: body weight ↓ 10–15%	2								
No food or water intake	3								
CLINICAL SIGNS									
Normal Temperature, Cardiac and Resp. rates	0								
Slight changes	1								
T ± 1 °C, C/R rates ↕ 30%	2								
T ± 2 °C, C/R rates ↕ 50% or very ↓	3								
NATURAL BEHAVIOUR									
Normal	0								
Minor changes	1								
Less mobile and alert, isolated	2								
Vocalization, self-mutilation, restless or very still	3								
PROVOKED BEHAVIOUR									
Normal	0								
Minor depression or exaggerated response	1								
Moderate change in expected behaviour	2								
Reacts violently, or very weak and precomatose	3								
SCORE ADJUSTMENT									
If you have scored a 3 more than once, score an extra point for each 3	2–5								
TOTAL	0–20								

JUDGEMENT

0–4 Normal 5–9 Monitor carefully, consider analgesics. 10–14 Suffering, provide relief, observe regularly. Seek second opinion from day-to-day care person and/or named veterinary surgeon. Consider termination. 15–20 Severe pain. Does your experimental protocol need rethinking?

11.1 Distress scoring sheet.

If there is no deviation from the normal, score zero. If there is mild deviation, score 1, moderate deviation score 2, and substantial deviation score 3.

If three is scored more than once, given an extra 1 to each, making a maximum score of 20.

Score 0–4   Normal

Score 5–9   Monitor carefully. Consider giving analgesics.

Score 10–14   The animal is suffering. Use analgesics and keep under observation.

Score 15–20   Is this experiment really worthwhile?

The use of such a system encourages regular close observation of the animal, leading to improved standards of animal care. If the animal is found to be deteriorating then actively consider termination before the end point is reached. Remember to re-score the animal **after** giving analgesics. When the level of pain and distress has been scored, records must be kept to ensure the procedure is kept within the severity banding allocated on the Project Licence. If this severity banding is exceeded the Home Office Inspector must be informed. For experiments where the animal's condition is expected to deteriorate the use of the distress scoring sheet is invaluable in fixing the humane end point to a certain limit when the animal must then be humanely destroyed. If any doubt exists then **the welfare of the animal must come first, and the responsibility for this lies with the personal licensee.**

SPECIES-SPECIFIC SIGNS OF PAIN AND STRESS
Mouse
Different strains of mice may vary significantly in their responses to particular stimuli but generally, an increase in sleeping time and weight loss will follow any procedure expected to cause pain. Affected animals show piloerection and hunched appearance and may be isolated from the rest of the group.

Rat
Rats are generally docile but become more aggressive, and resist handling during repeated stressful procedures. Acute pain or distress is accompanied by vocalization and struggling. They will lick or guard a painful area and will sit crouched. Sleep patterns will be disturbed and increased if pain and distress are present. The production of an ocular discharge is common, known as chromodacryorrhoea, as the animal produces haematoporphyrin-stained exudate with a reddish appearance.

Guinea-pig

Guinea-pigs are alert, apprehensive animals who will try to avoid capture and restraint. Any unusual sign of acceptance indicates the animal is unwell. Loud vocalization accompanies even minor and transient pain. They often appear sleepy when in pain and rarely show aggression. They are stoical animals and it can be difficult to assess whether they are in pain from a single glance. A carefully used pain scoring assessment method is needed.

Hamster

Hamsters will show weight loss, extended sleep period, and increased aggression or depression. Ocular discharge and diarrhoea may be associated with stress.

Rabbit

A rabbit will often react to painful procedures with stoic acceptance. This may relate to feral behaviour where concealment is important for survival. A rabbit in pain is usually characterized by reduced food and water intake and limited movement. There may be apparent photosensitivity and ocular discharge with protrusion of the third eyelid. Faecal staining of the coat, digestive disturbances, and dehydration may also be seen.

Cat

Cats in pain tend to be quiet and subdued and stop eating and grooming. Some react more violently, especially to acute and/or severe pain with frenzied activity or even dementia. Previously friendly cats may become aggressive when handled or approached if they are in pain. Vocalization is variable as cats in pain will generally remain silent until attempts are made to move them, when they may howl and growl. Purring is not always a sign of health and contentment in cats, as very sick cats frequently purr.

Dog

They are generally quieter and less alert than usual, with a 'hangdog' look. Stiffness or unwillingness to move may be noticed. Often there is inappetance, shivering, and panting. With less severe pain, dogs are often restless, but if in severe pain they tend to lie still and crouch. They may whimper or howl, and growl without provocation. They may bite and scratch at painful areas and may become more vicious or aggressive.

Ruminants

Sheep will appear dull and depressed with little interest in their surroundings. There is inappetance and weight loss and there may be rapid shallow respiration. Grunting and grinding of the teeth may be heard. Rumination and eructation may be reduced.

Primates

Monkeys will often appear to show little reaction to pain. They will have a generally miserable appearance, and may adopt a huddled position or crouch with head forwards and arms across the body. They have a 'sad' expression with glassy eyes. They may moan or grunt, and tend to avoid companions. Grooming may stop, and food and drink are usually refused. Ill monkeys may attract extra attention from cage mates, varying from social grooming to attack. Acute abdominal pain may be shown by facial contortions, clenching of teeth, restlessness, and shaking. Vocalization is more likely to indicate anger than pain.

12 Anaesthesia of laboratory animals

Successful anaesthesia does not depend simply on the types of drugs, doses, and routes of administration used. A good standard of **animal care** must also be maintained, both pre and post-operatively, including reduction in stress and provision of pain control. All these factors must be taken into account when designing an anaesthetic regime. If proper care is not taken during an anaesthetic, then recovery may be prolonged. Semi-conscious animals may lie in their urine or faeces and develop skin lesions, or get bedding in their eyes and noses. Their cage mates may attack them, and they may remain inappetant for long periods if recovery is slow. These and other problems such as post-operative pain are distressing for the animal, and can be avoided by good intra-operative and post-operative care.

REQUIREMENTS OF AN ANAESTHETIC

The primary reason for giving an anaesthetic is to provide **humane restraint**, a reasonable degree of **muscle relaxation** to facilitate procedures, and most importantly, sufficient **analgesia** to prevent the animal experiencing pain.

The effects of the anaesthetic must be **consistent and repeatable**.

The drugs must have a **wide safety margin**, both for the animal, and the user.

Good anaesthetic practice is also relevant to the **scientific validity** of any study using animals. Most animal models are carefully defined to have the smallest degree of unwanted variations. If the animal is to recover from experimental surgery, it must return to **physiological normality**, or a defined state of abnormality, as rapidly as possible. Pain, fear, distress, inappetance, hypothermia, hypoxia or respiratory acidosis, which may occur with poor anaesthesia, do not make for a good animal model. So **physiological stability** and **minimal deleterious effects** are required during the anaesthetic.

An anaesthetic regime should be selected which interferes as little as possible with the experiment, and either **does not alter**

the measurements being recorded, or alters them in a consistent manner which can be accounted for. This aim can be difficult to achieve, but a careful review of the agents available and their known physiological and pharmacological effects can minimize the interaction between anaesthetic and animal model.

Equipment

Care should be taken to choose equipment for pre-medication and anaesthesia which is suitable for the species of animal involved. **Most anaesthetic will pass easily through a 25 gauge needle and the smallest gauge practicable for the species should be used** (see chapter 10).

PRE-OPERATIVE CARE

Animals which are healthy and disease-free are less likely to have problems during anaesthesia than those with overt or subclinical disease. Those with respiratory infections in particular should be avoided. It is best to use animals of known health status (see Chapter 7), but this may not be possible, and sometimes a previous experimental procedure may have altered the health of the animal. Either way, animals should be given a **clinical examination** prior to anaesthesia to rule out infectious and non-infectious diseases. The thoroughness of the examination depends on the species, the procedure, and the likely duration of anaesthesia. As a minimum, the respiratory and cardiovascular systems should be examined. This can be achieved simply by careful observation of the animal noting the rate and character of the breathing and the colour of the extremities. Animals should be eating normal quantities of food and water, and show no obvious signs of disease.

Animals need to become acclimatized to their surroundings for 1 to 2 weeks before the examination, or normal physiological stress responses will result in abnormal experimental measurements (Manser 1992).

Adequate pre-operative **training** of animals to accept the procedure calmly, and sensible **individual animal selection** will reduce stress. The temperaments of individual animals vary and some will be more suited to certain types of procedure than others.

Fasting in rabbits and rodents prior to anaesthesia is not required, as they cannot vomit, and rodents rapidly become hypoglycaemic if deprived of food. They may be fasted for upper gastrointestinal tract surgery, but the stomach will only be empty if coprophagy is prevented.

Careful and **expert handling** of the animals is important. The fear and stress associated with movement from the animal holding room to the operating theatre should be considered. In many cases, using a **sedative** or **tranquillizer** to calm the animal before moving it from its home cage will reduce the stress during the induction period. Anaesthesia aims to provide humane restraint and an absence of pain, but unnecessary distress should not be caused while trying to achieve these aims.

Pre-anaesthetic medication
The aims of premedication are:

1. To decrease fear and apprehension, to aid in stress-free induction;
2. To reduce the amounts of other anaesthetic agents required to induce general anaesthesia, so reducing their undesirable side-effects;
3. To smooth recovery from anaesthesia;
4. To reduce salivary and bronchial secretions, and to block the vaso-vagal reflex, in which bradycardia occurs due to endotracheal intubation and handling of the viscera;
5. To reduce post-operative pain.

Pre-anaesthetic drugs
1. *Anticholinergics* These block parasympathetic stimulation and decrease salivary and bronchial secretions, which is especially important in smaller animals. They also protect the heart from vagal inhibition, which may occur when viscera are handled. The drug of choice is atropine sulphate, at a concentration of 600 μg per ml.

2. *Tranquillizers and sedatives.* Tranquillizers produce a calming effect without causing sedation. At high doses, they cause ataxia and depression but the animal is readily roused, especially in response to painful stimuli as these agents have no analgesic action. Sedatives produce drowsiness and appear to reduce fear and apprehension in animals. There is considerable species variation in the effect of these drugs and some overlap in the action of the agents, making definitive classification as sedative or tranquillizer difficult. However, all of these drugs potentiate the action of anaesthetics, hypnotics, and narcotic analgesics and are therefore useful in calming the animal before induction of anaesthesia and in reducing the dose of other drugs required to produce surgical anaesthesia.

   (a) *Phenothiazines.* e.g. acepromazine, chlorpromazine. All of these drugs cause hypotension and a fall in body temperature. Care should be taken with aged animals or those with impaired cardiovascular function.

(b) *Butyrophenones.* e.g. fluanisone. These are more potent than phenothiazines but less hypotensive. They are primarily used as a component of neuroleptanalgesia.

(c) *Benzodiazepines.* e.g. diazepam, midazolam. In some species (rabbits, rodents) these cause marked sedation, together with good skeletal muscle relaxation, and are potent anticonvulsants. They potentiate the action of many anaesthetics and narcotic analgesics and are a valuable adjunct to many anaesthetic regimes.

(d) *Xylazine* acts on α_2-adrenoceptors, and is a potent sedative in many species. Its major use is in combination with ketamine to produce surgical anaesthesia. It potentiates the action of other anaesthetics but is best not used with barbiturates or alphaxalone/alphadolone as the combination causes severe respiratory depression.

(e) *Medetomidine.* This is a specific α_2-agonist producing marked sedation. It has potent anaesthetic sparing properties. It is a relatively new drug, and its major advantage is that it has a specific α_2-antagonist, atipamezole, and is therefore rapidly reversed, enabling the animal to return to physiological normality much more rapidly.

(f) *Narcotic analgesics.* These have minimal sedative effects but reduce the amount of anaesthetic necessary to produce surgical anaesthesia and may provide some post-operative analgesia.

GENERAL ANAESTHESIA

The choice of an anaesthetic agent will depend on many factors. Most important is the humane treatment of the animal, and the safety of personnel involved in the procedure.

All agents used will affect the physiology of the animal to some extent. With **balanced anaesthesia**, drugs are administered in combinations, including premedicants, injectable and inhaled anaesthetics and analgesics, to achieve the best physiological stability of the animal, and to reduce the undesirable side-effects of some drugs. Anaesthesia is a combination of narcosis, relaxation, and analgesia. Different drugs induce these states to differing degrees.

It is likely that anaesthetic regimes of different durations and depths will be required for different procedures, and that some anaesthetic drugs will be contraindicated in certain protocols. Therefore, **advantage should be taken of the wide variety of anaesthetic drugs and combinations available**, rather than sticking to one or two methods which are unlikely to be best suited to all experimental protocols.

When choosing an anaesthetic regime, it must be remembered that surgical techniques and anaesthesia have effects on the physiology of the animal. Surgical procedures have potentially major disruptive effects which can be prolonged. The stress of surgery results in corticosteroid release, and levels can remain high for several days. The more invasive the surgery, the greater the stress response, which is designed to assist the animal in overcoming injury, but may be undesirable in the experimental subject, and unnecessary if peri-operative care is good.

The physiological effects of different anaesthetic regimes are quite minor compared with those produced by surgery, so the possibilities of short-term minor alterations in experimental results should not be used as an excuse for adhering to an unsuitable anaesthetic technique. The anaesthetic should be chosen when **all** factors potentially influencing the protocol have been considered.

Inhalation or injection?

This is often the first choice made by an anaesthetist. Rather than immediately discarding one group of techniques, the possibility of using both methods should be considered. Frequently, injectable agents are used for induction, and inhaled agents for maintenance of anaesthesia. The advantage of maintaining the animal on gas is that control of anaesthesia is much greater with this system. Inhaled agents are mainly eliminated by the lungs, whereas injectable agents need to be metabolized by the liver, and excreted by the kidneys. This process can be prolonged. Recovery is therefore more rapid from inhaled agents, which is important in regaining normal physiology, to control post-operative hypothermia and fluid or electrolyte imbalance. Many of the newer injectable anaesthetics have specific reversal agents, which speed recovery and overcome many of these potential problems. **It is better to have a rapid recovery and provide adequate post-operative analgesia, than to have prolonged anaesthesia.**

If injectable agents are used, it is sometimes presumed that an anaesthetic machine is not needed. **Oxygen supplementation** is therefore not given, and hypoxia may develop, particularly in long-term anaesthesia which lasts for more than one hour, in which **respiratory depression** can lead to hypercapnia and acidosis. Similarly, endotracheal intubation may only be considered if the animal is to be connected to an anaesthetic machine and ventilated. The ability to administer oxygen and control ventilation if required, helps control many potential difficulties encountered in anaesthesia.

Reversal of overdose of anaesthetic is quicker and usually more successful with inhalation agents than injectable ones, so reducing mortality.

Other factors influencing the choice of anaesthetic include the

species involved, the **length of the procedure**, the **depth** of anaesthesia required, and the **nature of the procedure**.

Inhaled agents

The most basic technique for anaesthetizing a small rodent is to place it in a jar containing a pad of cotton wool soaked in volatile anaesthetic. This technique has nothing to recommend it. Direct contact with the anaesthetic is very irritant to mucous membranes, and the concentration of anaesthetic in the jar is unpredictable. Potent anaesthetics such as halothane, which vaporize readily, build up to dangerously high concentrations. If ether is used there is considerable hazard of fire and explosion, and with any agent it is impossible to prevent contamination of the operating theatre environment, which presents a hazard for the anaesthetist and others.

Every laboratory should invest in an **anaesthetic machine**, to deliver known concentrations of anaesthetic to the animal, and construct an induction chamber. This should have transparent sides, for easy observation, and be easy to clean. Disposable tissue should be placed on the floor of the chamber, and the chamber should be thoroughly cleaned between animals, or the smell of the previous occupant could cause distress. Chambers are simple and convenient for small rodents, but should not be used for larger animals such as rabbits or cats.

If a commercial anaesthetic machine is too expensive, a simple apparatus can be constructed from readily available components, using a laboratory trolley as the basis. The required components are:

> **Cylinders of gas.** These are available in several sizes. Oxygen is delivered in black cylinders with white necks, nitrous oxide in blue cylinders, and carbon dioxide in grey cylinders.

> **A reducing valve** attaches to the gas cylinders to ensure that a constant low pressure of gas is delivered to the flowmeter.

> **The flowmeter** monitors the amount of gas flowing to the animal. It consists of a graduated glass tube with a bobbin, which floats at a level determined by the amount of gas passing through the tube. The tube is calibrated such that the **top** of the bobbin is in line with the marking indicating the flow rate of the gas. The bobbin should rotate in the tube. This rotation can be detected by observing a small spot which is painted in the centre of the bobbin. Each flowmeter is calibrated specifically for a particular gas.

> From the flowmeter, gas passes into a **vaporizer**, which contains the volatile anaesthetic. Oxygen is diverted

Table 12.1
Use of volatile agents in rodents and rabbits

Anaesthetic	Induction concentration (%)	Maintenance concentration (%)
Halothane	2–4	0.8–2
Methoxyflurane	4	0.4–1
Enflurane	3–5	1–3
Isoflurane	3–4	1.5–3

into the vaporizer to collect the vapour. The amount of oxygen passing into the vaporizer can be controlled, thus changing the concentration of anaesthetic delivered to the animal. The calibrations on each vaporizer are specific for a particular volatile agent, due to the differing physical properties of the agents.

Anaesthesia can be induced in a chamber or via a face mask (see Table 12.1 for suitable concentrations). Once the animal is anaesthetized, as judged by lack of movement and monitoring the rate and depth of breathing, it can be removed from the chamber. For most procedures, the volatile agent is then administered via an endotracheal tube or face mask at a suitable maintenance concentration for as long as required. For procedures lasting less than 1 minute, maintenance may not be required.

If the use of a face mask or chamber is resented by the animal, as commonly occurs with larger species, stress can be reduced by administering a sedative or tranquillizer in the home pen, 45 minutes before induction, and/or by using an injectable agent for induction and a volatile agent for maintenance.

Circuits
In order to maintain an animal with volatile anaesthetic, a system is required to deliver gases from the anaesthetic machine to the animal. Placing an impregnated pad over the animal's muzzle is unsatisfactory, as there is no control over the depth of anaesthesia, and there is pollution of the operating theatre environment. Use should be made of a suitable circuit. Examples are shown in Figures 12.1–12.5. The particular one chosen will depend on the size of the animal, the site of the operative field, whether intermittent positive pressure ventilation (IPPV) is to be used, and personal preference.

The quantity of oxygen piped into the circuit depends on the needs of the animal (which depends on its bodyweight), and the

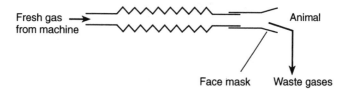

Disadvantages: IPPV not possible
Scavenging waste gases not possible

12.1 Basic open circuit.

type of circuit. For all except closed circuits, the expired air is not re-inhaled, i.e. rebreathing should not occur. If the oxygen flow rate in these open circuits is too low, then rebreathing may occur, and carbon dioxide can build up to dangerous levels as there is no method of removing it. It is essential therefore to have a high enough flow rate.

The concentration of volatile agent does **not** depend on the weight of the animal, but on the agent itself. Therefore, larger animals do not need a higher concentration of anaesthetic.

Face masks and endotracheal tubes
At the end of the tubing, a tight-fitting face mask can be used to deliver the anaesthetic. This must not rub the eyes. The disadvantages of this are that gas invariably leaks from around the mask,

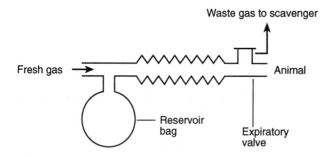

Advantages: Can be used for IPPV
Scavenging possible
Disadvantages: Valve causes resistance, so only suitable for animals over 5kg

12.2 Magill circuit.

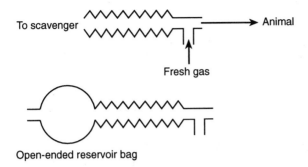

Open-ended reservoir bag

Advantages: Little resistance to breathing

Small dead space

Ventilation—by occluding the end of the tube, or by modification to include a bag

Gas flow needed is twice minute volume (respiratory rate × tidal volume)

12.3 Ayres T-piece.

contributing to pollution of the atmosphere, and ventilation (IPPV) is not possible should it be necessary.

Endotracheal intubation requires skill, and in some species the use of an appropriate laryngoscope, but does not have the disadvantages of the face mask.

Volatile anaesthetic agents are toxic to people. Prolonged exposure is thought to cause hepatotoxicity and abortion, and other side-effects. It is necessary to use some sort of **scavenger system**, for example:

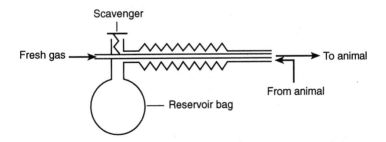

Flow rate of 200–300 ml/kg/min needed to prevent rebreathing

12.4 Bain coaxial circuit.

1. Administration in a fume cupboard. This has practical difficulties of access to the animal.
2. A length of tubing attached to the expiratory port can be used to duct waste gases to the outside, either actively or passively.
3. Using a fluosorber, which actively scavenges the gases using charcoal. Portable scavenging units are available from IMS (see Directory p. 274). Remember the canister must be changed when it is exhausted.

All anaesthetic equipment must be regularly serviced to ensure the correct quantities of vapour are delivered. Also the machine must be checked before each anaesthetic is given to ensure that:

1. Oxygen cylinders are full, spare cylinders are available, and a key to turn them on is handy.
2. Flowmeters are working.
3. The vaporizer is working and contains enough volatile agent.
4. The emergency oxygen flush is working.
5. A suitable circuit is correctly attached.
6. Endotracheal tubes are in good condition.
7. The scavenging system is attached and functional.
8. Any monitoring equipment is set up properly.

Only when all these have been checked should the animal be anaesthetized.

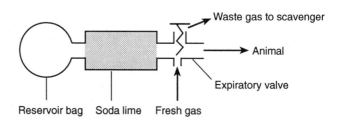

Advantages:	The soda lime absorbs CO_2, so that gas may be rebreathed from the reservoir bag.
	Small fresh gas flow rates are needed, which reduces the cost.
Disadvantages:	The valve and the soda lime increase resistance, and dead space, so only suitable for animals over 15 kg.

12.5 Closed circuit.

Volatile agents

Volatile agents are delivered to the animal from an anaesthetic machine using oxygen, or a mixture of oxygen and nitrous oxide, as the carrier. There are several volatile agents available.

Methoxyflurane produces a slow induction and slow recovery, making it very safe, with good analgesic activity, which lasts into the post-operative period. It can be used for both induction and maintenance. Its metabolism causes release of fluoride ions, which may cause renal damage, but this is only significant in specific studies, or in prolonged anaesthesia.

Halothane is very widely used. It is a non-flammable liquid, which is very potent, but has a high therapeutic index and is therefore very safe. It is non-irritant to the mucous membranes, so causes little increase in salivary and bronchial secretions. It has a cardiodepressant effect, reducing heart rate and blood pressure, and it sensitizes the heart to the arrhythmic effects of catecholamines. There tends to be some shivering on recovery. It is the cheapest of the modern volatile agents.

Enflurane has many similarities to halothane.

Isoflurane produces more respiratory depression than halothane, but it is not metabolized to toxic by-products so it is safer. It has little effect on hepatic enzyme levels which may be important in some studies. The maximum exposure limit for personnel is higher than for halothane, so it is often used for this reason, although it is considerably more expensive.

Desflurane (1653) and *Sevoflurane* are relatively newly developed volatile agents.

All of these agents produce smooth induction. With methoxyflurane, induction is slower, but this is only a problem if it is used with a face mask in larger species, e.g. cats or rabbits, for inducing anaesthesia.

Halothane and methoxyflurane induce liver microsomal enzyme systems, which isoflurane does not, making it suitable for studies requiring maintenance of normal drug metabolizing ability.

Whichever agent is chosen, it may be used in conjunction with **nitrous oxide gas**. Nitrous oxide cannot be used for induction of anaesthesia in most species, and is therefore not suitable for use alone or with neuromuscular blocking agents. However, it has minimal cardiovascular and respiratory effects, so if used with a volatile agent it reduces the required concentration of that agent, reducing its side-effects. It is used 60:40 or 50:50 with oxygen to deliver the volatile agent. After prolonged anaesthesia, 100 per cent oxygen should be given for 5 to 10 minutes, or oxygen can be displaced from the lungs by the nitrous oxide as it is breathed off, causing diffusion hypoxia which can lead to suffocation. Nitrous oxide is not absorbed by charcoal, and so scavenging systems do not remove it from the waste gases.

Nitrous oxide reacts with vitamin B_{12}, causing depletion of this vitamin after anaesthesia lasting more than 6 hours.

Ether has been widely used, but has many problems associated with its use. It is very irritant to mucous membranes making induction unpleasant for the animal, they frequently cough and hold their breath, and it can exacerbate pre-existing respiratory disease. Induction and recovery are relatively slow, making inadvertent overdose less likely, but prolonging the period of recovery. There are periods of voluntary and involuntary excitement before anaesthesia is induced. Ether stimulates the sympathetic nervous system, increasing circulating catecholamine levels and causing hyperglycaemia. It also induces liver microsomal enzyme systems. There is good muscle relaxation with ether anaesthesia, and it is cheap. However, it is twice as heavy as air, so vapour tends to pool on the floor, and unless ventilation is good it can easily explode if there is a spark. It can even be ignited if there is a spark in an adjacent room.

The explosive nature of ether makes it a serious safety hazard, and in its traditional role, the ether jar, **it should be replaced with methoxyflurane**.

Provided that anaesthesia has not prolonged, **recovery** is usually **rapid**, but if animals have been maintained on volatile agents for 1 hour or more, full recovery can take more than 15 to 20 minutes. The speed of recovery varies depending on the attention which has been given to **controlling the depth of anaesthesia** during surgery, by varying the concentration of agent delivered to the animal. Generally, as anaesthesia progresses, the concentration of anaesthetic can be reduced slightly. After major surgery, a further reduction in depth can be made during suturing of subcutaneous tissues and skin, so reducing recovery time following completion of surgery. This ability to vary the depth of anaesthesia is one of the major advantages of using inhalation agents.

Injectable agents
General considerations
Injectable agents are given either intravenously, so they go directly into the bloodstream, or via another parenteral route from which they are absorbed into the blood. They are particularly useful for the induction of anaesthesia, which can then be maintained by inhalation, or for short duration anaesthesia. Intravenous injections usually act very rapidly, producing loss of consciousness immediately. If given by other routes these agents should ideally produce anaesthesia in the time it takes for the blood to circulate from the injection site to the brain just once. This makes it easier to titrate the dose to the exact requirements of the animal. With intravenous

injections, the calculated dose is administered slowly, allowing time for the drug to circulate before deciding whether more is required. The drug can therefore be administered to effect.

Some of the injectable agents have a slower onset of action, such as ketamine, or chloral hydrate, which has to be metabolized to an active compound before having any effect. These drugs are more difficult to use as induction agents, as time is taken before effects are seen and it is difficult to judge if more is required.

There are difficulties associated with the use of injectable agents for the induction of anaesthesia. A considerable amount of skill is required to perform intravenous injections, particularly in small rodents, and it takes time and practice to acquire these skills. For injectable agents more than inhaled agents, the level of anaesthesia achieved depends on the level of stimulation of the animal at the time. A animal which is given an injection and left undisturbed may be apparently deeply anaesthetized, but once surgery starts it may respond by reflex movements of the limbs and an increase in the depth and rate of breathing. This does not necessarily mean that a further injection of anaesthetic is required: if more is given, then surgery stops and the animal is no longer stimulated, there is a risk that the level of anaesthesia will become dangerously deep. This phenomenon also occurs with inhaled agents, but to a lesser degree. **It is therefore important to be familiar with the techniques required for the administration of injectable agents, and the type of anaesthesia obtained, and to develop the necessary skills for monitoring accurately the depth of anaesthesia**.

Recovery

With inhaled agents, if an overdose is given, the vaporizer can simply be switched off and the animal allowed to breathe oxygen for a few minutes until the anaesthetic gas has been exhaled. With injectable agents, once they are given, they have to be metabolized by the liver and excreted by the kidneys. This process can be prolonged, and recovery from injectable anaesthetics usually takes longer than for inhaled agents. This increases the risk of hypothermia or dehydration developing. With some of the barbiturate drugs, return to consciousness is rapid because the drug is redistributed away from the CNS into other body tissues such as fat, due to their high lipid solubility. However, much drug is still present in the body and there is the potential for the animal to relapse into unconsciousness if the drug recirculates into the CNS, which can occur in animals with little body fat. Some injectable drugs have specific reversal agents, the administration of which will lead to a return to consciousness within minutes. The reversal agents can also be used to reverse any respiratory depression which develops in the post-operative period. Some injectable agents are 'ultra-short-acting', and are metabolized very quickly. Use of these newer drugs has overcome some of the problems previously associated with injectable agents.

Injectable agents in laboratory species

The small size of many of the laboratory species and the problems of restraint in others means that injectable agents are most frequently given via the intramuscular or intraperitoneal routes. These routes require a **large total dose** of the drug to be given. Drugs are usually administered as a single bolus, so it is essential to **weigh** the animal first, as a 'guesstimate' of the weight is likely to be wildly inaccurate and result in over- or under-dosing the animal. It is important therefore that drugs used in laboratory species have a wide safety margin, are non-irritant, and can be administered in a small volume through a narrow gauge needle (25 or 27 gauge—see Chapter 10).

Absorption from intraperitoneal, subcutaneous or intramuscular injections can be slow, as it depends on the blood flow to the site, so induction can be slow. Recovery is also slow, and the residual effects of the drug can persist for long periods. There is also a time lag between the injection of the drug and deepening of anaesthesia, so if these agents are used for long procedures, good monitoring of the depth of anaesthesia is essential, in order that top-up injections can be given in good time. **It is not acceptable to give a further dose by injection as the animal starts to wake up**. Intravenous injections allow dosing to be adjusted according to the individual animal's response, so there is less likelihood of over- or under-dosing. They also have many advantages in terms of control of depth of anaesthesia, and also more drugs can be given by this route than many others. It is worthwhile developing the skill required for giving intravenous injections. They can be given using needles and syringes, butterfly needles, or over-the-needle cannulae, which can be maintained in a vein for long periods. To facilitate venepuncture in nervous animals, local anaesthetic cream can be applied 30–60 minutes beforehand, to desensitize the skin (EMLA cream, Astra).

Choice of anaesthetic agent

The quality of anaesthesia produced by the different agents varies considerably. Many of the injectable agents produce poor analgesia, and are insufficient for major surgery. The addition of a low level of a volatile agent can improve the level of analgesia. Therefore, before choosing an anaesthetic regime, the nature of the procedure should be considered.

Many injectable agents are available.

Barbiturates

Pentobarbitone (e.g. Sagatal) is a barbiturate which depresses the CNS, and produces marked cardiovascular and respiratory depression. It is weakly analgesic, and therefore very large doses must be given before pain perception is reduced. This means that it has a low therapeutic index: the lethal dose is only slightly above the clinical dose. After intravenous (i.v.) injections, pentobarbitone takes a long time to cross the blood-brain barrier into the CNS, so onset of anaesthesia is slow. Injections must therefore be given very slowly, to allow anaesthesia to deepen to its full extent before giving extra doses. If it is given carefully, safety can be improved, but if the injection is intraperitoneal (i.p.) and given as a single bolus, it has a poor safety margin, and mortality can be high. Recovery from pentobarbitone anaesthesia is slow, and may be associated with convulsive movements and paddling. Pentobarbitone is best used at low doses to induce anaesthesia, with an inhaled agent used to provide analgesia, or for terminal anaesthesia. It is often used as it is cheap. In concentrated form, pentobarbitone is used for euthanasia.

Thiopentone is a thiobarbiturate, which must be injected i.v., as it is very irritant if given perivascularly. It acts rapidly after injection, inducing anaesthesia almost immediately, and should be given carefully to avoid over-dosage. It takes several days to be broken down by the liver, but anaesthesia is of short duration because thiopentone is absorbed by the body fat, and taken away from the CNS. Therefore, the duration and depth of anaesthesia produced with thiopentone depends on the amount of drug injected, the speed of the injection, and the rate of absorption by the body fat. It should be used with extreme care in animals with low body fat levels. It is not suitable for repeated injections for maintenance, as it builds up in the body causing prolonged anaesthesia. It tends to produce apnoea on induction, muscle relaxation is poor, and it is poorly analgesic. To improve analgesia, thiopentone should be given after premedication with a suitable analgesic, and anaesthesia supplemented with an inhaled agent or opioid analgesic.

Methohexitone is a potent barbiturate which is short-acting. Induction is smooth and rapid, but recovery tends to be associated with muscle tremors and convulsions.

Thiamylal and *Inactin* are barbiturates which are similar to thiopentone. They are irritant and should not be given i.p.

Non-barbiturates

Propofol (e.g. Rapinovet) is a substituted phenol, which is administered intravenously. It acts rapidly, inducing anaesthesia smoothly without excitatory side-effects. It is ultra-short-acting, and recovery is rapid and smooth. It can be used by continuous infusion for prolonged anaesthesia. It can be used safely in rats, cats, dogs, monkeys, pigs, and rabbits, and can be combined with a wide range of premedicants, analgesics, and inhaled agents.

Saffan is a mixture of steroids, alphaxalone and alphadolone.

It was first used as an anaesthetic for cats, but is now used in almost all domestic species except dogs. There is a component in the mixture (cremophor EL) which causes histamine release in dogs, and can cause anaphylaxis. Saffan is given i.v. in cats, rats, rabbits, sheep, and primates. It is the agent of choice for general anaesthesia in all primates (after pre-medication with ketamine in Old World monkeys), and is used for sedation in New World monkeys. In rabbits however, analgesia is poor, and these are best given an alternative agent. Intramuscular and intraperitoneal injections produce variable results, and large volumes of the drug are required, which limits its use in these sites.

Intravenous injection rapidly results in anaesthesia, but induction is not as smooth as with thiopentone. In cats, muscle relaxation is good, but there is some depression of the CNS, cardiovascular, and respiratory centres. There may be some trembling, paddling, or even convulsions on recovery, and oedema of the paws and ears may develop. Overall, however, it has five features which make it popular as an anaesthetic agent:

1. It has a high therapeutic index.
2. It does not accumulate in the body.
3. Recovery of consciousness and appetite is rapid after administration ceases.
4. There is little respiratory depression but muscle relaxation is good.
5. It is not irritant if given outside a vein.

Saffan can be used for short-term anaesthesia, for induction and maintenance with an inhaled agent, or for long-term anaesthesia by intermittent injections or continuous infusion.

Dissociative agents
Ketamine can be given by intramuscular or intravenous injection. It circulates rapidly and produces analgesia but little muscle relaxation. There is an apparent lack of awareness of the surroundings, and analgesia varies with the species. The corneal reflex is lost very early with ketamine, and in long procedures it is necessary to protect the eye with a bland ophthalmic ointment. The laryngeal and pharyngeal reflexes are maintained, but there is an increase in salivary secretions which could cause airway obstruction, so it is often used with a drying agent such as atropine. There is good maintenance of blood pressure under ketamine sedation. In rodents, the dose required to produce light surgical anaesthesia with ketamine alone causes severe respiratory depression. It is generally insufficient for surgery when given alone, but it can be combined with xylazine, medetomidine, diazepam, midazolam, or saffan to produce general anaesthesia in

many species (hamster, rabbit, cat, primate, pig, sheep). It is the agent of choice for sedation of Old World monkeys. Using ketamine in combination with other agents also avoids the side-effects such as muscle tremors.

Narcotic analgesics (opioid analgesics)

These are potent analgesics, but they may produce respiratory depression as a side-effect. They may be pure agonists (morphine like-activity), or partial agonists, producing mixed activity, with analgesia and some respiratory depression combined with some antagonistic activity. Pure antagonists exist for these drugs, which can be used to reverse their effects.

Fentanyl is a potent analgesic which lasts 15-20 minutes. In rats, dogs, and primates it has sedative effects, but in mice, cats and horses it causes excitement. Respiratory depression can be marked, and facilities for IPPV should be available. It is useful in neuroleptanalgesia, or as a component of balanced anaesthetic regimes.

Alfentanil is less potent than fentanyl, but the onset of activity is more rapid. Respiratory depression is marked. It is used in combinations, to increase analgesia in animals anaesthetized with other agents, or as a premedicant to reduce the amount of induction agent required.

Sufentanil is ten times more potent than fentanyl, but has not yet been extensively used in animals.

Neuroleptanalgesics

These combine a neuroleptic (tranquillizer) with a narcotic analgesic, to suppress some of the side-effects of the narcotic. Alone, neuroleptanalgesics produce some respiratory depression, and poor muscle relaxation, but when combined with a benzodiazepine these side-effects are markedly reduced. So, combinations of fentanyl/ fluanisone (Hypnorm) with midazolam (Hypnovel), or diazepam (Valium) can be used to provide good anaesthesia for rodents and rabbits. Midazolam is particularly useful since it is water-soluble and can therefore be administered in the same syringe as the Hypnorm. It can also be diluted with sterile water for accurate dosing in small rodents.

The actions of neuroleptanalgesics can be reversed by the use of narcotic antagonists, such as naloxone, or partial agonists, such as **buprenorphine** or **nalbuphine**, which will also provide post-operative analgesia. Naloxone reverses analgesia as well as any side-effects such as respiratory depression, so it must not be used unless surgery has been completed. Doxapram, which is a respiratory stimulant, may be more useful, as it reverses the respiratory depression without affecting the neuroleptanalgesia, as it has no pharmacological antagonistic activity against the opioid. Reversal is not always required, as most neuroleptanalgesics will wear off by themselves, but anaesthesia will be prolonged.

Medetomidine and atipamezole

Medetomidine (Domitor) is a specific an α_2-agonist, and is reversed by its specific antagonist, atipamezole (Antisedan). Medetomidine can be given by intravenous, intramuscular, or subcutaneous injection. When used as a sedative, the animal must be left undisturbed for 10-15 minutes after administration for the drug to reach maximum effect, preferably in a quiet environment, as they may still react to noise.

Medetomidine does not induce general anaesthesia when used alone, but it has great potential for use in combinations. It reduces the requirement for halothane by about 70 per cent and can be used with fentanyl/fluanisone to produce good anaesthesia and analgesia. It may also be used successfully with ketamine in many species.

The great advantage of medetomidine is that it can be reversed very rapidly with atipamezole, and if it is used in combinations with neuroleptanalgesics, buprenorphine or nalbuphine can be used as a partial antagonist for fentanyl, combining rapid recovery with the provision of analgesia.

Medetomidine can cause heavy urination, so care must be taken to prevent the animal becoming wet, hypothermic or even dehydrated. Also there is a fall in blood pressure, and cyanosis may be apparent, with venous blood being very dark. Despite this, blood oxygen saturation remains high.

Other drugs

Metomidate (Hypnodil) is an hypnotic agent used for the anaesthesia of pigs, horses, and birds. It produces good muscle relaxation, but poor analgesia, so is usually used in combination with other agents, such as fentanyl or azaperone. If given intravenously after premedication with azaperone, induction is smooth and there are few side-effects. There are few respiratory or cardiovascular effects. However, it is not suitable for major surgical procedures without the addition of an analgesic.

Etomidate is similar to metomidate. It has been used in rats. It is potent, and produces anaesthesia, the duration of which is dose-dependant. If it is injected intravenously, there may be involuntary movements, but these can be reduced by the use of a premedicant. It appears to be safe and produces little or no cardiovascular depression.

Chloral hydrate is an hypnotic agent which produces anaesthesia at high dose rates with cardiovascular and respiratory depression. Anaesthesia is slow in onset, as the chloral hydrate has to be metabolized to trichloroethanol, which is the active compound. Chloral hydrate is no longer considered to be an anaesthetic agent. It has been used in combination with pentobarbitone and magnesium sulphate as an anaesthetic for the horse (Equithesin). If given to rats, paralytic ileus may develop post-operatively.

Alpha-chloralose is reputed to have minimal effects on cardiovascular and respiratory systems at low dose rates. Large volumes of solution have to be given before consciousness is lost, and it is often given after induction by another agent. The anaesthesia produced is light, and muscle twitches are common. It is slow in onset, and recovery is slow. Its use is restricted to long-lasting light anaesthesia for procedures involving a minimum of surgical interference, and is best used for terminal procedures.

Urethane has similar characteristics to alpha-chloralose but is a better analgesic. However, it is carcinogenic.

Tribromoethanol (Avertin) can be given i.p. but decomposition of stored solutions can cause severe irritation and adhesions. Administration of a second anaesthetic at a later date is associated with high mortality.

Non-recovery anaesthesia

The same pre- and intra-operative care should be given to animals undergoing terminal procedures as to those which will recover. In fact, as terminal procedures are more likely to involve invasive techniques the animal may require more care. Some anaesthetics are only suitable for non-recovery procedures, e.g. pentobarbitone, which is poorly analgesic and associated with high mortality; alpha-chloralose, which has a very prolonged recovery period; and combinations which include chloral hydrate, which can cause paralytic ileus post- operatively in the rat.

Long-term anaesthesia

In choosing an agent for long-term anaesthesia, there are four options. A **single dose** of long-acting anaesthetic can be given, such as alpha-chloralose. This produces a light plane of anaesthesia, but induction is slow and associated with excitement. It is preferable to induce anaesthesia with a short-acting anaesthetic or a volatile agent, and then administer the alpha-chloralose. There is considerable variation in the depth of anaesthesia produced. In the dog, surgical anaesthesia is probably not achieved.

Long-term anaesthesia can be successfully achieved with **intermittent injections**. Inevitably, the plane of anaesthesia varies markedly. If the 'top-up' is not administered in time, the animal may experience pain. **Good anaesthetic monitoring is essential.** One of the best methods is to administer an anaesthetic intravenously by continuous infusion, e.g. Saffan, propofol, fentanyl/midazolam, or etomidate/

fentanyl. Short-acting agents enable the plasma concentration and depth of anaesthesia to be adjusted quickly, and recovery is likely to be rapid.

Long-term anaesthesia can also be achieved by the use of inhaled agents for maintenance, after induction with inhaled or injectable agents.

Combination techniques

Using several anaesthetics together can overcome many of the disadvantages encountered when using the agents individually. For example, administering sedatives prior to induction with an inhaled agent, or inducing anaesthesia with an injectable agent, reduces the stress otherwise caused by induction with inhaled agents.

Some injectable agents produce little analgesia alone, and addition of a low concentration of an inhaled agent is a safe way of providing this, especially in birds.

The aim of these **balanced anaesthetic** regimens is to minimize the interference to the animals physiology caused by the drugs, and to enable recovery to be smooth, rapid, and pain-free.

Local anaesthetics

Another option, especially in farm species, is to use local anaesthetics. These can be used to infiltrate around the area of proposed surgery, or they can be used to block specific nerves to the surgical field. If the anaesthetic is given by the epidural or subdural route, the area of anaesthesia can be quite extensive. However, in order to produce adequate restraint for procedures and to reduce stress, it is usually preferable to use general anaesthetics.

Whatever method of anaesthesia is selected, high standards of animal care are essential, if meaningful research data are to be obtained and animal welfare is to be maintained. For doses of anaesthetic agents, see Chapter 13.

ANAESTHETIC MANAGEMENT

All anaesthetic drugs will, if given in sufficient quantity, cause death. On the other hand, if insufficient is given, the animal will feel pain. To prevent these two extremes from occurring, care must be taken in the maintenance of **respiratory function**, **circulatory function**, and **body temperature**, and in monitoring the **depth of anaesthesia**.

Depth of anaesthesia

Anaesthesia cannot be described simply as awake, asleep or dead. Historically, it has been described as having four stages, with stage three (surgical anaesthesia) being divided into three planes. This

classification was devised in man for use with a single volatile anaesthetic, and relies on assessment of cardiovascular and respiratory
signs. In animals, with different species and for balanced anaesthetic
regimes using combinations of drugs, this becomes unworkable.

The animal should be **monitored frequently**, at least every 5
minutes, or more often if stability has not been established, so
that the anaesthetist knows exactly what depth of anaesthesia the
animal has reached. More than one sign should be monitored. The
depth cannot be judged from one sign alone, in isolation from the
rest of the animal. The different combinations of drugs used and the
nature of the procedure should be borne in mind, as these will affect
the depth of anaesthesia.

It must be ensured that sufficient depth of analgesia has been
reached to prevent the animal perceiving pain, and that consciousness has been lost to a sufficient degree to prevent distress. This
can only be judged by the presence or absence of certain reflexes,
and by assessment of vital functions. The pedal withdrawal reflex
is the most commonly used reflex, also pinching the tail or ear. The
responses are abolished if surgical anaesthesia has been reached,
but they give no indication if the anaesthesia is too deep. In
this state, the animal may be in danger of dying from respiratory
and cardiovascular failure. The signs are development of a fixed,
staring eye, slow shallow breathing, or deep gasping breaths, blue
colouration of the mucous membranes, and a fall in blood pressure.
**Monitoring of the vital signs will detect these changes early on, and
enable action to be taken to prevent deterioration.**

Respiration
This is easily monitored, and assessments should be made of the
rate, **depth**, and **pattern** of breathing. This is done by watching the
movement of the chest wall, or reservoir bag, or in larger animals
by the use of an oesophageal stethoscope. It is wise to use an
apnoea alarm, even if using a mechanical ventilator, as these can
be disconnected easily.

The effectiveness of pulmonary gas exchange can be assessed by
the **colour of the mucous membranes**, and of blood being shed
at the site of surgery. Virtually all anaesthetics cause some respiratory depression, leading to hypoxia and hypercapnia. Hypoxia
may be indicated if the colour of blood or mucous membranes
changes, but hypercapnia is not. Providing the animal with oxygen
intra-operatively helps to prevent the development of hypoxia.
Carbon dioxide concentration can be assessed by measuring the
end tidal carbon dioxide, or by direct arterial blood gas analysis.
Transcutaneous oxygen and carbon dioxide monitoring equipment
can be used in some species.

The respiratory rate usually falls during anaesthesia, and a fall of up to 50 per cent from the normal is acceptable. If the rate continues to drop, there may be a problem. If the animal is not breathing, it is not necessarily too deeply anaesthetized. It may be too light, and holding its breath, which demonstrates the need to monitor more than one variable.

If there is genuine apnoea, first ensure that the **airway is unobstructed**. The oropharynx may be blocked with mucus or blood, or the endotracheal tube may be kinked. Make sure that the thoracic movements are **not restricted** by the position of the animal, or by limbs being restrained too tightly. Check that undue pressure is not being placed on the chest wall by over-enthusiastic use of retractors, or by inadvertent leaning on the animal.

Respiration can be stimulated by moving the endotracheal tube, artificial ventilation, needle stimulation of the philtrum, or by respiratory stimulants such as doxapram. Neuroleptanalgesic combinations can be reversed with naloxone, and medetomidine can be reversed with atipamezole, but only if surgery has been completed.

In most instances, it is advantageous for the animal to be intubated, or in non-recovery cases for a tracheostomy to be performed, to facilitate IPPV. Even if IPPV is not required, this enables a clear airway to be maintained.

Cardiovascular signs

The simplest method of monitoring is to assess the rate rhythm and quality of the **pulse**, using a superficial artery depending on the species. An oesophageal stethoscope will indicate heart rate and rhythm, and **capillary refill time** indicates tissue perfusion. This can be tested by applying pressure briefly to a mucous membrane to blanch it, then measuring how long it takes for the pink colour to return. It should take only two to three seconds.

Cardiac failure may occur dramatically as cardiac arrest, but more usually it is gradual in onset, and monitoring can prevent a disaster. Some anaesthetics cause a fall in cardiac output, and hypotension. Hypoxia and hypercapnia, resulting from respiratory depression, cause cardiac arrhythmias. Loss of blood or body fluids can cause hypovolaemic shock and cardiac arrest. Blood loss commonly causes death in small rodents because a small loss represents a high proportion of the animal's total blood volume. This problem can be minimized if careful attention is paid to surgical technique. Use of a blood donor is potentially useful. Transfusion reactions are rarely encountered with an initial transfusion, and never if an animal of the same inbred strain is used.

A secure **venous line** should be established for infusion of fluids for circulatory support. This allows easy administration of analgesics and anaesthetics, and facilitates dosing with stimulant drugs in case of an emergency.

An **ECG** will give an indication of cardiac function, but it only indicates the electrical activity of the heart. It is possible to have a normal ECG with a cardiac output of zero.

Body temperature

It is essential to monitor this in small animals. All anaesthetics affect thermoregulation, and an animal's body temperature will fall unless measures are taken to prevent this. The fall is exacerbated by the flow of cold air from the anaesthetic machine, shaving the animal, use of cold skin preparations, placing on a cold operating table, exposing viscera during surgery, and administering cold fluids. Smaller animals have a larger surface area to volume ratio, so are particularly susceptible to heat loss. Animals less than 1 kg in weight will require extra heating to avoid a drop in body temperature.

Temperature can be monitored using a rectal or oesophageal probe, and core temperature can be compared with skin surface temperature. The difference should be less than 2–3°C. Alternatively, the temperature can be monitored by simply feeling the animal's paws and ears.

Heat loss can be minimized by insulation, with cotton wool, Vetbed, foil or bubble packing. Additional heating can be provided with heat lamps or heat blankets (see Chapter 16). Care must be taken not to burn the animal, and a thermostatically controlled heating pad is the best solution.

If fluids are to be given, these should be warmed first. A bag of intravenous fluid can easily be warmed to body temperature by immersing it in warm water, or using a Safe and Warm instant heat pouch (see Chapter 16).

It is important to ensure that measures to prevent hypothermia are continued throughout the recovery period. This can be achieved in small animals by using incubators. For adults, the temperature should be 25–30 °C, and for neonates 35–37 °C. Bedding also helps provide insulation, e.g. Vetbed, tissue paper. Sawdust should not be used as this will stick to the animal's nose and mouth, or to wound surfaces.

Hypothermia is the commonest cause of mortality in small rodents, so monitoring of body temperature and taking steps to prevent hypothermia are vitally important.

Ocular signs

These are less useful as indicators of the depth of anaesthesia. The palpebral reflex is lost at variable times depending on the species. In rodents it is hard to assess, and in rabbits it may not be lost until anaesthesia is very deep. If ketamine is used, the reflex is abolished very early.

Observation of the position of the eyeball, the degree of pupillary dilatation, and whether nystagmus is occurring are all useful indicators once experience is gained with one species and a particular anaesthetic regime.

Aside from monitoring vital signs, it is useful to monitor the **continued function of the anaesthetic delivery system**. Infusion pumps and anaesthetic machines should be fitted with alarms which sound when the machines fail, or are empty. Also personnel carrying out prolonged procedures may become fatigued, and steps should be taken to prevent this.

ANAESTHETIC EMERGENCIES

If an animal goes into circulatory arrest, during or after an anaesthetic, and the cerebral circulation is not restored within three minutes, irreversible brain damage will occur.

In order to maintain tissue oxygen levels, blood must firstly contain sufficient oxygen, and secondly, circulate properly to and from the tissues. Therefore, monitoring and resuscitation attempts are directed towards:

- **A** for **Airway**,
- **B** for **Breathing**, and
- **C** for **Circulation**, in that order.

A First, ensure that the animal has a patent **airway**, with no respiratory obstruction. Hold the head and neck straight, draw the tongue forwards, and remove any mucus from the mouth and throat. If an endotracheal tube is used, ensure it is not kinked or blocked with mucus. Also check the anaesthetic delivery circuit for blockages or empty oxygen cylinders.

B Secondly, make sure the animal is **breathing**. Respiratory arrest is frequently preceeded by respiratory insufficiency which has been undetected. Once the airway is clear (as in A), oxygen should be administered. Artificial ventilation may be manual or mechanical, or by chest compression. The pulse may also be checked once tissue oxygen levels have improved (pink colour returning to tongue or conjunctiva). Then the cause of apnoea can be investigated. DO NOT DELAY GIVING OXYGEN. EXCESS WILL DO NO HARM: INSUFFICIENCY IS FATAL.

Causes of apnoea
Common causes

- Airway obstruction
- Resistance to breathing in anaesthetic circuit
- Central depression due to:

– Anaesthetic overdose
– Hypoxia and hypercapnia, e.g. empty oxygen cylinder
• Light anaesthesia—animals often hold their breath

Less common causes

• Hypocapnia from hyperventilation (encountered if IPPV used)
• Mechanical prevention of breathing, e.g. too many instruments or leaning on chest wall
• Neuromuscular blocking agents
• Pain (especially thoracic)
• High spinal or epidural block

Respiration may be stimulated in many ways:

– Stimulation of nasal septum with 21 gauge needle—acts via acupuncture point
– Reversal of anaesthetic agent
– Use of respiratory stimulant, e.g. doxapram (Dopram)— 1–2 mg/kg after injectable anaesthetics or 5–10 mg/kg for inhaled anaesthetics, i.v., or with sublingual drops
– Control of thoracic pain

C   Thirdly, check the **circulation**. There may be:

Inadequate circulating volume

Most anaesthetics are hypotensive, causing vasodilation, and also depress many of the physiological mechanisms which would normally maintain blood pressure in the event of blood loss or other circulatory problem.

Cardiac massage should be given if no pulse is felt. If the animal is hypovolaemic and has gone into shock (most cases), then cardiac massage will be ineffective unless fluids are also administered. These should be warmed first.

Cardiac deficiency

• Cardiac arrhythmias will usually recover when respiration is adequate.
• Bradycardia may be reversed with atropine given subcutaneously.
• In the event of cardiac failure, it should be determined whether there is ventricular asystole, or ventricular fibrillation.
• For ventricular asystole, adrenaline is given.

- For ventricular fibrillation, lignocaine is given to cause asystole, then massage is continued and adrenaline administered.
- After resuscitation, animals may need treatment for acidosis, with sodium bicarbonate; and cerebral oedema, with corticosteroids or diuretics.

Hypothermia

The development of hypothermia should be considered whenever recovery from anaesthetic is slow, particularly in small animals in which it is a serious problem.

Causes

- Reduced heat production
- Increased heat loss

Little can be done to increase heat production. To reduce heat loss, the fur should not be wetted excessively and dry drapes should be used to minimize evaporative losses. The environment should be warm, between 20 and 22°C, and ideally the operating table should be heated. A water blanket kept at 38°C is best.

Fluids to be administered should first be heated to 38°C.

Shivering to increase heat production results in an increase in oxygen demand, therefore oxygen should be given to the hypothermic animal.

Anaesthetic emergency checklist

In the event of cardiorespiratory failure:

1. Note time. Remember 3 minute emergency
2. A—Airway. Clear and maintain airway.
3. B—Breathing. Ventilate with oxygen 20 times per minute. Use doxapram if required.
4. C—Circulation. Cardiac massage.

Then:

Give *fluids* to restore the circulation (warmed first).

Restore *cardiac rhythm*
- For asystole give adrenaline, at 1 ml/kg of 1:10,000 solution i.v.
- Repeat after 10 minutes if necessary.
- For fibrillation give lignocaine, 1–2 mg/kg by intra-cardiac injection.

Once the animal is stabilized continue ventilation and support the circulation (fluids, adrenaline).

Watch for *hypothermia*.

Treat for *acidosis* or *cerebral oedema* if required. Bicarbonate can be given at 1.5 ml/kg of 5 per cent solution slowly i.v. for acidosis. Corticosteriods such as methylprednisolone (SoluMedrone V) at 20–30 mg/kg slow i.v. injection, or dexamethasone at 2–4 mg/kg i.v. can be given to reduce cerebral oedema, which can also be controlled with a diuretic such as frusemide (Lasix) at 0.5–5 mg/kg i.v. or i.m.

13 Post-operative care and analgesia

....

When considering prevention of pain, it may not be necessary to reach immediately for analgesic drugs. Consider first whether the standard of post-operative care is as high as it can be, as this can contribute enormously to the prevention of pain occurring.

INTRODUCTION

All of the parameters monitored during surgery (see Chapter 12) should continue to be monitored in the immediate post-operative period. Some special attention will be required post-operatively, so it is preferable to have a specific recovery area where individual nursing care may be given. Emergency drugs and equipment should be available (see Chapter 12).

Warmth and comfort should continue to be provided.

Respiratory depression frequently develops post-operatively but it often goes unnoticed, resulting in death. Respiration must be monitored, and treatments given as required. If respiratory depression continues, a continuous infusion of the drug doxapram may be needed.

Fluid balance must be maintained. It is vitally important to support the circulation by correcting fluid imbalances.

Blood loss may occur gradually during surgery. Swabs can be weighed to estimate the loss, but blood also seeps into drapes and body cavities, and this blood loss may pass unnoticed.

Plasma loss occurs by exudation from tissues and into the peritoneal cavity during prolonged abdominal surgery.

The extracellular fluid is depleted by evaporation from the respiratory tract, exposed viscera, and wounds.

In addition to these losses, the animal may not drink for 12–24 hours post-operatively. **Fluids must be given** to replace the losses and to cover the period when intake is low. Fluid requirements are **40–80 ml/kg/24 hours**, and this is best given orally if the animal is conscious and can swallow. Depending on the species, the fluid can be offered as a drink or administered orally via syringe, gavage tube or stomach tube. Alternatively, moist food can be offered, such as mashes to large animals or Transgel (Charles River Ltd) or reconstituted fruit jelly to rodents. If not, subcutaneous or intraperitoneal administration of isotonic saline or dextrose saline (0.18 per cent saline with 4 per cent dextrose) can be substituted. As a guide, fluid should be replaced at 10–15 ml/kg. If dehydration is severe, as judged by skin tenting and loss of skin tone, fluids must be given intravenously.

In a normal animal, loss of 15–20 per cent of circulating blood volume will cause signs of hypovolaemic shock. In an anaesthetized animal, many of the mechanisms which normally maintain physiological stability are depressed, so losses which are less severe than 15 per cent can have serious consequences.

Monitoring *urine output* helps give an indication of the animals hydration status, although retrospectively. Reduced urine output may be due to dehydration, urinary tract injury, or pain. In larger species, if the bladder is distended it should be catheterized and emptied, if it has not emptied spontaneously, to improve comfort post-operatively.

If the animal fails to *defaecate*, this may be due to absence of faeces, or paralytic ileus, which occurs if the bowels are handled excessively during surgery or with certain anaesthetic agents (e.g. chloral hydrate in the rat). Careful surgical and anaesthetic techniques are important to avoid this. Sometimes the animal is constipated, and this may be corrected by the administration of a suitable enema.

Monitoring of *bodyweight* or individual food and water consumption helps provide an assessment of recovery from surgery (see Chapter 11). For cats and dogs, attention should be given to the provision of a suitable diet post-operatively. It should be appetizing but not so rich as to predispose to diarrhoea.

The *animal's position* must be considered carefully. If the animal is restrained in one position for a prolonged time, ensure that ties are not too tight, and that bony areas are padded to prevent pressure sores. A larger animal which is recumbent post-operatively for a prolonged period must be turned and moved regularly to prevent nerve or muscle damage. If the animal is on one side for too long, hypostatic pneumonia may develop. Ruminal tympany in ruminants can be avoided by passage of a stomach tube.

In animals anaesthetized for a long period, and those in which ketamine has been used, the *cornea* must be prevented from drying by the use of a bland ophthalmic ointment.

Good sterility and surgical technique (see Chapter 14) will go a long way towards reducing post-operative problems, and reducing the need for analgesic drugs.

The amount of *individual attention* given to the animal during the post-operative period depends on the species. Companion animals react well to personal contact, whereas rodents or rabbits may be stressed by it. The design of the recovery area should take account of the species and their individual needs. The *noise level, light intensity, and temperature* should be appropriate. The ambient light level should be fairly low for the animal's comfort, but capable of being raised to a brighter level in order to be able to examine the animal satisfactorily. The temperature should be higher than usual in order to prevent the development of post-operative hypothermia. Animals prefer a familiar environment, and may or may not prefer the presence of *other animals*, depending on how they were housed before surgery. Companionship must be balanced with the risk of bullying and cannibalism. The *caging and bedding* must provide warmth and comfort, and keep the

PPL no.:   Licensee: Severity limit:

Animal ID: Date:

Nature of operation:

Anaesthetic details:

Post-operative drugs:

Date Time	Recorded by	Urine	Faeces	Food	Water	Drugs administered	Animal observations/ comments	Stress score

13.1 Post-operative record.

animal clean and dry. Appropriate use of well-applied bandages and pressure pads can help control pain in limbs and around the head (see Jones' Animal Nursing). *Fluid therapy* should be given as outlined above.

The animal must be frequently observed, and the findings recorded, together with any drugs administered. The record chart should be easily available. An example of a post-operative record chart is shown in Figure 13.1 and it can be adapted to the individual needs of the procedure.

MANAGEMENT OF PAIN

Under the 1986 Act, 'Pain, suffering, distress, and lasting harm' are to be interpreted in their widest sense, to include death, disease, injury, physiological and psychological stress, significant discomfort, or any disturbance to normal health.

Pain and suffering can be largely controlled by appropriate treatments, so pain is unnecessary in the majority of scientific procedures. All animals should receive post-operative pain assessment, and analgesics administered if they are needed. Following their administration, the animal must be reassessed to ensure that the pain has been adequately controlled. On some occasions, one type of analgesic may be contraindicated, but it is most unlikely that no suitable analgesic will be available.

It is important to remember that pain produces physiological changes that alter the rate of recovery from surgical procedures, and these changes may affect the experiment as well as the animal's welfare. The assessment of pain is discussed in Chapter 11.

As well as considering whether pain may be present, it is also important to consider whether the animal might be suffering 'distress' (see Chapter 11). A cold wet environment, without any suitable bedding material is likely to cause distress to many animal species, and states of physiological imbalance, such as dehydration caused by inadequate fluid therapy, would not be referred to as painful, but could cause distress. It is important to recognize that pain can have an emotional component, and in man, both the intensity of pain as reported by the patient, and the requirement for analgesics to control pain are increased by factors such as fear and apprehension.

Pain can be alleviated by the systemic use of centrally or peripherally acting analgesics, by the use of local anaesthetics, and in the larger animals, by the application of supporting bandages to protect and immobilize damaged tissues. The particular type of treatment chosen will depend on the species, the nature of the pain, its cause, and its estimated severity and duration, but whatever treatment is chosen the aim will be to reduce the discomfort to the animal as much as possible.

Experience in animals and controlled trials in man have shown that analgesics are most effective in controlling post-operative pain if they are administered *before* pain is experienced. It is therefore preferable to administer opioids either as part of the pre-anaesthetic medication, or intra-operatively before pain is perceived by the animal, as part of the **balanced anaesthetic regime** (see Chapter 12).

In man, it has been demonstrated that more effective pain relief is provided by the technique of administering opioids by continuous infusion rather than by intermittent dosing. This approach can be used in animals by adding opioids to intravenous fluids administered via a giving-set, but there can be practical difficulties. If analgesics have been given as part of the balanced anaesthetic regime, they will control pain before it is experienced, and if extra doses are required they may be administered by constant intravenous infusion, subcutaneous or intramuscular injection, or orally. If the animal is eating and drinking normally they may be administered in food which reduces the disturbance caused by giving injections, and any associated stress caused to the animal. Rats will take analgesic drugs mixed into a cube of jelly and it has been found that the preferred flavours are the berry fruits (Pekow 1992).

USE OF OPIOIDS TO CONTROL PAIN

If pain is assessed to be moderate or severe, opioids are the drugs usually required to produce pain relief. A wide variety of different opioids are available, but the duration of action of most of them is under 4 hours. However, buprenorphine has been shown to have a longer duration of action of up to 12 hours in some species.

At present, buprenorphine is usually the opioid analgesic of choice in laboratory animal species.

Some researchers have expressed concern about the wide range of effects that opioids have, which are unrelated to their analgesic action. These potential side-effects should not, however, be used as an excuse for withholding pain relief. The most serious consequence of over-dose with opioids in man is respiratory depression. This may occasionally be seen in animals given very high doses of morphine or pethidine, or if potent agonists such as fentanyl or alfentanil are administered, and can be alleviated by the use of doxapram. Significant respiratory depression rarely occurs following the use of mixed agonist/antagonist drugs such as buprenorphine, nalbuphine, and butorphanol.

THE USE OF NON-STEROIDAL ANTI-INFLAMMATORY DRUGS (NSAIDs)

These compounds are often used in the management of mild or moderate pain in animals, and in circumstances when the use of opioids is contraindicated. Available compounds may be classified as carboxylic or enolic acids. See Figure 13.2 for their classification. The NSAIDs are anti-pyretic and analgesic through central actions, and anti-inflammatory, analgesic, anti-thrombotic,

Carboxylic acids	Enolic acids
Salicylates	*Pyrazolones*
Sodium salicylate	Phenylbutazone
Acetylsalicylic acid	Oxyphenbutazone
Quinolines	Dipyrone
Cincophen	Isopyrin
Aminonicotinic acids	*Oxicams*
Flunixin	Piroxicam
Clonixin	Meloxicam
Propionic acids	Tenoxicam
Naproxen	
Ibuprofen	
Carprofen	
Diclofenac	
Ketoprofen	
Suprofen	
Anthranilic acids	
Meclofenamic acid	
Mefenamic acid	
Tolfenamic acid	
Indolines	
Indomethacin	

13.2 Classification of non-steroidal anti-inflammatory drugs. (From Lees *et al.* 1991.)

and anti-endotoxaemic through peripheral actions. As well as therapeutic activity, they also have some undesirable side-effects which include gastrointestinal ulceration, nephrotoxicity, hepato-toxicity, blood dyscrasias, urticaria, and teratogenic effects. Many factors including drug formulation, age, diet, and **stress** affect the predisposition to the various side-effects.

The newer NSAIDs, such as carprofen, ketoprofen and flunixin, appear to be useful alternatives to opioids in the control of even quite severe post-operative pain. These are long-lasting and carprofen does not produce gastrointestinal ulceration, unlike some of the other non-steroidals. Ketoprofen is 15 times more potent than phenylbutazone and 30 times more potent than aspirin. Following intravenous injection, there is activity within 2 hours which reaches a peak at 12 hours and is still present at 24 hours. Care must be taken with the use of flunixin during the immediate post-operative period or during surgery due to potential nephrotoxicity. The use of NSAIDs in the dog and cat has been reviewed by Taylor (1985), and in laboratory animals by Liles and Flecknell (1992).

LOCAL ANAESTHETICS

Local anaesthesia is relatively under-used in laboratory animals. The use of agents such as xylocaine, lignocaine or bupivicaine should also be considered for controlling pain. They may be applied topically, infiltrated locally around a wound, or infiltrated around a major sensory nerve supplying a specific area of the body for regional anaesthesia. In some circumstances, this can provide a good method of post-operative pain control.

ADDITIONAL CONSIDERATIONS IN PAIN MANAGEMENT

The use of analgesics should be included not only with the anaes-thetic regime but also with the overall plan of the animal's care. It has been shown in man that the provision of effective analgesia will reduce the time taken for post-operative recovery. Surgery should be planned for a time in the day and a time in the week when there will be adequate numbers of staff available to provide an adequate level of observation of the animal. Good post-operative care is essential for the animal's welfare and for good scientific practice and the responsibility for that care of the animal lies with the licensee, since in the standard conditions under which personal licences are granted 'the personal licensee must take effective precautions, including the use of sedatives, analgesics or anaesthetics, to prevent or reduce to the minimum level consistent with the aims of the procedure any pain, suffering distress or discomfort in the animals used.' (Home Office Guidance notes, Appendix VI.)

DRUG DOSES FOR VARIOUS SPECIES
MOUSE
Combinations of recommended drugs
Drug doses are intended only as a guide and may have to be altered to take account of varying responses of different strains of mice to the drugs.

Hypnorm (fentanyl/fluanisone) combinations. If sedation alone is required, with some analgesia but with poor muscle relaxation, Hypnorm can be administered alone at 0.5 ml/kg.

For surgical anaesthesia lasting 45–60 minutes use 0.2 ml/kg Hypnorm mixed with 1 mg/kg midazolam i.m. It is best to make up a 1:10 dilution of the bottled solutions and use this for accurate dosing. A 50 g mouse will need 0.1 ml of each i.m.; give the Hypnorm first then the midazolam once it is sedated. Alternatively a mixture of 1 part Hypnorm: 2 parts sterile water for injection: 1 part midazolam can be given at the rate of 10 ml/kg i.p. The fentanyl component can be reversed with naloxone (0.01–0.1 mg/kg i.v., i.m. or i.p.) to reduce the recovery time.

Diazepam may be used instead of midazolam (0.4 ml/kg Hypnorm plus 5 mg/kg diazepam) but they may not be mixed in the same syringe since the diazepam, unlike midazolam, is not water-soluble.

Fentanyl/medetomidine combinations in the mouse are not recommended since there can be urinary retention leading to rupture of the bladder.

Ketamine combinations. Ketamine alone, at a dose of 200 mg/kg, will produce sedation and immobilization for about 20 minutes.

For surgical anaesthesia ketamine at 75 mg/kg i.p. with medetomidine at 0.5 mg/kg i.p. will produce 20–30 minutes of surgical anaesthesia, with rapid recovery if atipamezole (1 mg/kg s.c., i.m. or i.p.) is used as a reversal agent, or 2–4 hours sleep time if it is not. Ketamine (100 mg/kg) can also be given with xylazine (10 mg/kg) for similar results although the atipamezole is not so specific in its action to reverse the xylazine.

Ketamine at 100 mg/kg i.m. with diazepam at 5 mg/kg i.p., or acetylpromazine at 2.5 mg/kg i.p., produces light anaesthesia for 20 minutes with 2 hours of sleep time.

Saffan. Alphaxalone/alphadolone at 10–15 mg/kg will produce 5 minutes surgical anaesthesia, but only if given i.v. Recovery will take about 10 minutes.

Propofol. Given i.v. at 26 mg/kg it may be useful for short-term anaesthesia or for induction prior to inhalation anaesthesia.

Barbiturates. Pentobarbitone has a narrow safety margin. If administered at 45 mg/kg i.p. it will produce anaesthesia lasting between 15 minutes and 1 hour. However, there will be marked respiratory depression and the animal will not be fully recovered for several hours. It is best used only for terminal procedures.

Thiopentone can be used i.v. at 30 mg/kg and methohexitone i.v. at 10 mg/kg.

Other drugs. Doxapram may be administered at 5–10 mg/kg i.m., i.v. or i.p. to reverse respiratory depression and atropine at 0.05 mg/kg will reduce salivary and bronchial secretions.

Analgesics
Opioids. *Buprenorphine*. At present the drug of choice for analgesia is buprenorphine up to 2.0 mg/kg s.c. given 12 hourly. It is a partial μ-agonist so reverses opioids such as fentanyl but maintains analgesia. It has a slow onset of about 40 minutes so must be given before the animal regains consciousness and feels pain, but it has a relatively long duration of action.

Butorphanol	Give 1–5 mg/kg s.c. for 4 hours analgesia
Codeine	Give 60–90 mg/kg orally or 20 mg/kg s.c. for 4 hours analgesia
Morphine	Give 2–5 mg/kg s.c. for 2–4 hours analgesia
Nalbuphine	Give 4–8 mg/kg i.m. for 3 hours analgesia
Pentazocine	Give 10 mg/kg s.c. for 3–4 hours analgesia
Pethidine	Give 10–20 mg/kg s.c. or i.m. for 2–3 hours analgesia

NSAIDs

Flunixin	2.5 mg/kg s.c. lasts 12 hours
Ibuprofen	30 mg/kg orally lasts 4 hours
Diclofenac	8 mg/kg orally
Paracetamol	200 mg/kg orally lasts 4 hours
Aspirin	120 mg/kg orally lasts 4 hours
Phenylbutazone	30 mg/kg orally

RAT
Combinations of recommended drugs
Medetomidine + Ketamine. Inject i.p. at a dose rate of 250–500 μg/kg medetomidine + 60–75 mg/kg ketamine. The medetomidine can be diluted 1:10 with sterile water and the ketamine added in the same syringe. Anaesthesia is achieved in a few minutes and

lasts up to 45 minutes. For recovery in 5–10 minutes, reverse with atipamezole 1 mg/kg s.c.

Hypnorm (fentanyl/fluanisone) combinations. If sedation only is required for minor procedures, use Hypnorm alone 0.2–0.5 ml/kg i.m. or 0.3–0.6 ml/kg i.p. and then reverse with naloxone 0.1 mg/kg.

Surgical anaesthesia with good muscle relaxation can be achieved for 40–60 minutes by making up a mixture of 1 part midazolam + 1 part Hypnorm + 2 parts sterile water and administering 2.7 ml/kg of the mixture i.p.

Alternatively, for better absorption, administer 0.3 ml/kg Hypnorm + 2.0 mg/kg midazolam i.m.

The fentanyl component of Hypnorm may be reversed by naloxone 0.1 mg/kg i.p., i.m. or i.v. or buprenorphine 0.05 mg/kg s.c.

A dose rate of 300 μg/kg fentanyl + 300 μg/kg medetomidine i.p. will provide 45–60 minutes of surgical anaesthesia. Reverse with atipamezole 1 mg/kg s.c. plus nalbuphine 2 mg/kg s.c. for recovery in about 8 minutes with post-operative analgesia.

Hypnorm may be given with diazepam (0.3 ml/kg i.m. Hypnorm plus 2.5 mg/kg i.m. diazepam) for 45–60 minutes of surgical anaesthesia.

Ketamine combinations. 10 mg/kg xylazine + 90 mg/kg ketamine i.p. will achieve about 30 minutes of surgical anaesthesia. Ketamine (100 mg/kg) alone may be used just for sedation but does not produce surgical anaesthesia. Ketamine at 75 mg/kg may also be combined with acetylpromazine at 2.5 mg/kg or with diazepam at 5 mg/kg to produce a light level of anaesthesia lasting 20 minutes with a 2–3 hour recovery time.

Saffan. Alphaxalone/alphadolone (Saffan): 10–15 mg/kg i.v. (lateral tail vein) will produce surgical anaesthesia for about 5 minutes but incremental doses may easily be given to prolong the anaesthesia.

Propofol. Given i.v. at 10 mg/kg propofol is useful for short duration anaesthesia or for induction prior to inhalation anaesthesia.

Barbiturates. Pentobarbitone has a narrow safety margin. If administered at 45 mg/kg i.p. it will produce anaesthesia lasting between 15 minutes and 1 hour. However, there will be marked respiratory depression and the animal will not be fully recovered for several hours. It is best used for terminal procedures.

Thiopentone can be used i.v. at 30 mg/kg and methohexitone i.v. at 10 mg/kg.

Non–recovery, long–term anaesthesia. Chloralose can be used at 55–65 mg/kg i.p. or urethane at 1000 mg/kg and will last 6–8 hours. Note that urethane is carcinogenic.

Other drugs. Doxapram may be administered at 5–10 mg/kg i.m., i.v. or i.p. to reverse respiratory depression and atropine at 0.05 mg/kg will reduce salivary and bronchial secretions.

Analgesics
Opioids. Buprenorphine: At present the opioid of choice for analgesia is buprenorphine at 0.05–0.5 mg/kg s.c. given 8–12 hourly. It is a partial μ-agonist so reverses opioids such as fentanyl but maintains analgesia. It has a slow onset of about 40 minutes so must be given before the animal regains consciousness and feels pain, but it has a relatively long duration of action.

Butorphanol	Give 2.0 mg/kg s.c. for 4 hours analgesia
Codeine	Give 60 mg/kg s.c. for 4 hours analgesia
Morphine	Give 2–5 mg/kg s.c. for 2–4 hours analgesia
Nalbuphine	Give 1–2 mg/kg i.m. for 3 hours analgesia
Pentazocine	Give 10 mg/kg s.c. for 3–4 hours analgesia
Pethidine	Give 10–20 mg/kg s.c. or i.m. for 2–3 hours analgesia

NSAIDs
Flunixin	2.5 mg/kg s.c. lasts 12 hours
Carprofen	10 mg/kg given orally lasts 18 hours
Ketoprofen	2 mg/kg once daily
Phenylbutazone	20 mg/kg orally
Diclofenac	10 mg/kg orally
Aspirin	100 mg/kg orally lasts 4 hours
Ibuprofen	15 mg/kg orally lasts 4 hours
Paracetamol	100–300 mg/kg orally lasts 4 hours

If the rat is eating, then continued analgesia may be provided by mixing the required amount of the drug in jelly so it is taken orally. The berry fruit flavours are preferred and the jelly also provides a source of fluids and glucose which aids in post-operative recovery (see Pekow 1992).

GUINEA-PIG

These are probably the most difficult rodents in which to achieve safe and effective general anaesthesia. The response to injectable agents is variable and post-anaesthetic complications such as respiratory infection, generalized depression, inappetance, and digestive disturbances may frequently be seen. Many of these problems may be avoided by careful selection of anaesthetic agents and high standards of pre-, intra-, and post-operative care.

The administration of atropine (0.05 mg/kg s.c.) is important to decrease airway obstruction in this species. To counteract respiratory depression give doxapram 5–15 mg/kg i.m. The injection of both Hypnorm and ketamine in the guinea-pig have been associated with tissue necrosis at the site leading to self-mutilation post-operatively.

Sedation
For sedation alone use Hypnorm (fentanyl/fluanisone) at 0.6 ml/kg i.m. but there will be poor muscle relaxation. Alternatively, use diazepam 2.5 mg/kg i.m. or acetylpromazine 2.5 mg/kg i.m. (hypotensive), or ketamine (100 mg/kg).

Injectable general anaesthesia
Fentanyl/fluanisone combinations. For surgical anaesthesia, combine Hypnorm 0.5 ml/kg with midazolam 2.5 mg/kg i.m. The fentanyl component can be reversed with naloxone (0.1 mg/kg) or, for continued analgesia with buprenorphine (0.05 mg/kg).

If combining Hypnorm with diazepam use 1 ml/kg Hypnorm plus 2.5 mg/kg diazepam. Fentanyl 160 μg/kg and medetomidine 400 μg/kg i.p. will give about 20 minutes anaesthesia in the guinea-pig but it can be of rather variable depth. Nalbuphine (1–2 mg/kg) and atipamezole (1 mg/kg) can be used for reversal.

Ketamine combinations. Xylazine and ketamine will give about 30 minutes of surgical anaesthesia (5 mg/kg xylazine s.c. plus 40–60 mg/kg ketamine i.m.). Ketamine may also be combined with diazepam (100 mg/kg ketamine plus 5 mg/kg diazepam i.m.), or with acetylpromazine (125 mg/kg ketamine plus 5 mg/kg acetylpromazine), or with medetomidine (40 mg/kg ketamine plus 250 μg/kg medetomidine). All these combinations give about 30 minutes anaesthesia with 2–3 hours recovery time. The response to the medetomidine component can be unpredictable in the guinea pig and is reversed with atipamezole (1 mg/kg).

Saffan. A short period of surgical anaesthesia can be induced intravenously (ear vein) with alphaxalone/alphadolone (Saffan) at the rate of 40 mg/kg.

Barbiturates. Pentobarbitone will induce anaesthesia lasting 15–60 minutes at 35 mg/kg i.p. but it has a very narrow safety margin. Methohexitone at 31 mg/kg i.v. can be used for induction.

Non-recovery procedures. Urethane will provide anaesthesia for 6–8 hours at 1500 mg/kg i.p. but it is carcinogenic and must be handled with care.

Analgesics
Opioids.

Buprenorphine	0.05 mg/kg s.c. lasts about 8 hours
Morphine	2–5 mg/kg s.c. or i.m. lasts 4 hours
Pethidine	10–20 mg/kg lasts 2–3 hours

NSAIDs

Ibuprofen	10 mg/kg i.m. lasts 4 hours
Diclofenac	2 mg/kg orally
Phenylbutazone	40 mg/kg orally
Aspirin	85 mg/kg orally lasts 4 hours

HAMSTER AND GERBIL
Fentanyl/fluanisone combinations. For sedation alone, use Hypnorm at 0.5–1 ml/kg i.m. For improved muscle relaxation or for surgical procedures, combine this with midazolam at a dose of 5 mg/kg i.m. It is best to give the Hypnorm i.p. and follow it up with the midazolam when the animal is quiet. Alternatively, mix up one part Hypnorm with one part midazolam (5 mg/ml) with two parts sterile water and use 4 ml/kg of the mixture i.p. Remember absorption from the peritoneal cavity is rather more variable than from the muscle.

The Hypnorm may be reversed with naloxone 0.05 mg/kg or for continued analgesia with buprenorphine 0.5 mg/kg.

Fentanyl plus metomidate (Hypnodil) can be used at a dose of 0.05 mg/kg fentanyl s.c. and 50 mg/kg metomidate s.c. in the gerbil to produce surgical anaesthesia.

Ketamine combinations. Ketamine 200 mg/kg plus xylazine 10 mg/kg i.p. will also produce surgical anaesthesia in the hamster but with moderate respiratory depression. The anaesthesia produced is not so reliable in the gerbil. A reduced dose of 2 mg/kg xylazine should be used, or ketamine at 50 mg/kg i.m. combined with diazepam at 5 mg/kg i.p.

Saffan. Alphaxalone/alphadolone (Saffan) may be given at 150 mg/kg i.p.

Medetomidine. Medetomidine at 100 μg/kg s.c. produces sedation which will enable cheek pouch examination. Higher doses do not seem to produce more sedation, but medetomidine is useful as a premedicant to reduce stress when inducing gaseous anaesthesia. To reverse the medetomidine use atipamezole at 1 mg/kg s.c.

Barbiturates. A dose rate of 50–90 mg/kg pentobarbitone in the hamster and 60–80 mg/kg in the gerbil will produce anaesthesia but with a high mortality rate as the safety margin is very low.

Other drugs. It is advisable to also use atropine 0.04 mg/kg s.c. when the animal is asleep to prevent secretions blocking the airway. For respiratory depression use doxapram 5–10 mg/kg.

Analgesics
Buprenorphine 0.5 mg/kg s.c. in the hamster, 0.1–0.2 mg/kg s.c. in the gerbil. Pethidine 20 mg/kg s.c.

RABBIT

In this species the combination of stress and general anaesthesia can lead to cardiac and respiratory arrest. It is therefore very important to keep both the pre-and post-operative stresses to the absolute minimum.

If there is respiratory depression or cyanosis, doxapram may be given at 2–5 mg/kg i.v.

Sedation
1. For sedation use *medetomidine* 250 μg/kg s.c. (i.e. about 0.75 ml Domitor for a 3 kg rabbit). Reverse with *atipamezole* 50 μg/kg into the ear vein.
2. *Acetylpromazine* 1 mg/kg will produce moderate sedation but is hypotensive.
3. *Diazepam* or *midazolam* 1–2 mg/kg i.v. or i.m. produces quite good sedation.
4. *Hypnorm* 0.22 ml/kg i.m. produces sedation and analgesia but poor muscle relaxation which lasts for about 20 minutes with full recovery taking more than 1 hour unless specific reversal with *naloxone* (0.1 mg/kg i.m.) is used.
5. *Ketamine* alone at 50 mg/kg i.m. produces sedation and immobilization lasting about 30 minutes.
6. *Xylazine* alone at 3 mg/kg i.m. will produce heavy sedation with some analgesia.

Injectable anaesthesia

1. *Hypnorm* 0.3 ml/kg i.m. followed by *midazolam* or *diazepam* at 2 mg/kg i.v. will give about 40 minutes of surgical anaesthesia with good muscle relaxation. It may be reversed with *naloxone* 0.1 mg/kg, or for continued analgesia with *buprenorphine* 0.02–0.05 mg/kg or *nalbuphine* 1–2 mg/kg i.v.

2. *Alphaxalone/alphadolone* (Saffan) 6–9 mg/kg i.v. will produce a light plane of anaesthesia but higher doses may be needed for surgery which can lead to sudden apnoea and cardiac arrest. It is good for long-term light anaesthesia or for a smooth induction followed by gaseous anaesthesia for maintenance.

3. *Medetomidine* can be combined with *fentanyl* for full anaesthesia. 330 μg/kg medetomidine plus 8 μg/kg fentanyl i.v. will give about 40 minutes anaesthesia and can be reveresed with 1–2 mg/kg *atipamezole* plus 1 mg/kg *nalbuphine*. Medetomidine can also be combined with ketamine (see below).

4. *Ketamine* (35 mg/kg i.m.) can be combined with *medetomidine* (0.5 mg/kg i.m.) or with xylazine (5 mg/kg i.m.) to produce surgical anaesthesia lasting about 30 minutes. Combined with *acetylpromazine* (50 mg/kg ketamine and 1 mg/kg acetylpromazine i.m.) or with *diazepam* or *midazolam* (25 mg/kg ketamine and 5 mg/kg of the benzodiazepine i.m.) a light plane of anaesthesia is produced.

5. *Propofol* given i.v. at 10 mg/kg will produce short-term anaesthesia lasting 5 minutes but is a useful induction agent prior to using gaseous anaesthesia. Similarly, *thiopentone* 30 mg/kg i.v. will produce short-term anaesthesia, as will *methohexitone* at 10 mg/kg i.v.

Non-recovery anaesthesia

1. *Pentobarbitone* at 30–45 mg/kg i.v. will induce surgical anaesthesia but has a very narrow safety margin. It is best used only for short-term non-recovery procedures.

2. *Chloralose* at 80–100 mg/kg i.v. will produce 8–10 hours of anaesthesia for long-term non-recovery procedures.

3. *Urethane* at 1000 mg/kg i.p. or i.v. will produce 6–8 hours of anaesthesia but it is carcinogenic.

Analgesics
Opioids.

Buprenorphine   0.01–0.05 mg/kg s.c. or i.v. will last for 8–12 hours

Butorphanol	0.1–0.5 mg/kg i.v. will last for 4 hours
Morphine	2–5 mg/kg s.c. or i.m. will last 2–4 hours
Nalbuphine	1–2 mg/kg i.v. will last 4–5 hours
Pentazocine	5 mg/kg i.v. will last 2–4 hours
Pethidine	10 mg/kg s.c. or i.m. lasts 2–3 hours

NSAIDs

Flunixin	1.1 mg/kg s.c. given twice daily
Ketoprofen	1.0 mg/kg i.m.
Aspirin	100 mg/kg orally will give about 4 hours relief from mild pain
Ibuprofen	10 mg/kg i.v.

DOG

Pre-anaesthetic preparation

A period of at least 12 hours starvation should precede anaesthesia in the dog. Premedication is essential to ensure a smooth induction and recovery from anaesthesia.

Sedation/Premedication

1. *Acetylpromazine* 0.2 mg/kg i.m. produces sedation but is not analgesic. It is markedly hypotensive.

2. *Opioids*
   – *Buprenorphine* (0.009 mg/kg) with acetylpromazine (0.07 mg/kg) i.m. produces moderate sedation suitable for minor procedures.
   – *Fentanyl-fluanisone* (Hypnorm) (0.1–0.2 ml/kg) or *fentanyl-droperidol* (Thalamonal) (0.1–0.15 ml/kg) i.m. produce heavy sedation and analgesia, but also cause bradycardia, which may be reversed with atropine.
   – *Alfentanil* (5–10 μg/kg) slowly i.v. prior to an induction agent will markedly reduce the dose of anaesthetic required.
   – *Papaveretum* (0.2 mg/kg) i.v. produces sedation.

3. *Xylazine* (2.0 mg/kg) i.m. produces good sedation with mild analgesia. It often produces vomiting, bradycardia, heart block, and hyperglycaemia. Heart effects are abolished by atropine.

4. *Atropine* (0.05 mg/kg) s.c. should be given prior to Hypnorm, alfentanil or xylazine, and in routine premedication prior to general anaesthesia.

5. *Medetomidine* produces sedation and analgesia which are

dose related. It can be given s.c., i.m. or i.v. Maximum effect is reached in 10–15 minutes. Blood pressure falls and breathing may become irregular, with cyanosis, but pO_2 is maintained.

The effects of *medetomidine doses* are as follows:

Dose	*Effects*
0.1–0.3 ml Domitor/10 kg	Slight sedation
0.3–0.8 ml Domitor/10 kg	Profound sedation and analgesia
0.1–0.2 ml Domitor/10 kg	For pre-anaesthesia

Atipamezole reverses the effects of medetomidine and xylazine if given 15–60 minutes after the sedative, i.m. or s.c.

To reverse medetomidine:

give an equal volume of atipamezole (if using the trade names Domitor and Antisedan which contain the drugs at 1 mg/ml and 5 mg/ml respectively)

To reverse xylazine:

Atipamezole at 200 μg/kg reverses the effects of 3 mg/kg xylazine

General anaesthesia

Injectable agents. Intravenous injection is straightforward using the cephalic vein on the anterior surface of the forelimb (see Chapter 9, Handling).

1. *Thiopentone* (Intraval). 10–20 mg/kg of 1.25–2.5% solution. Care must be taken to ensure none of drug goes perivascularly, since it is highly alkaline and causes tissue to slough.

2. *Methohexitone* (Brietal). 4–8 mg/kg of 1 per cent solution. Administer half the dose rapidly, then slowly to effect. Surgical anaesthesia last 5–10 minutes, and recovery takes 15–20 minutes. Without premedication, recovery is violent and analgesia poor. Respiratory depression often results in apnoea on induction.

3. *Propofol* (Rapinovet or Diprivan). 5–7.5 mg/kg i.v. This is suitable for induction or maintenance by continuous infusion.

   Note. The doses of injectable agents are markedly reduced after correct premedication (see Figure 13.3).

Inhalation agents. These should not be used for induction as the animals usually resent the procedure. They are usually used for maintenance after induction with injectable agents and endotracheal intubation.

Halothane, methoxyflurane, isoflurane and *enflurane* all produce stable anaesthesia with good analgesia and muscle relaxation. Methoxyflurane produces a slow, smooth induction and recovery, with a long period of post-operative analgesia.

Analgesia

Opioids. If combined with barbiturates, these are particularly respiratory depressant.

Buprenorphine	produces the longest relief, and may be used pre-, intra- or post-operatively. 0.01 mg/kg s.c., i.m. or i.v. lasts 6–8 hours.
Pethidine	this should be given i.m. 10 mg/kg post-operatively. Lasts 2–3 hours
Fentanyl	sometimes given i.v. during surgery to increase analgesia. 0.001–0.007 mg/kg
Butorphanol	0.4 mg/kg s.c. or i.m. 3–4 hourly
Nalbuphine	0.5–2.0 mg/kg s.c. or i.m. 3–8 hourly
Pentazocine	2 mg/kg i.m. 4 hourly
Morphine	0.5–5.0 mg/kg s.c. or i.m. 4 hourly
Codeine	0.25–5 mg/kg orally. Give 6 hourly in combination with paracetamol

Premedicant (i.m.)	Induction agent (i.v.)	
	Thiopentone mg/kg	Propofol mg/kg
No premedicant	10	6.7
Acetylpromazine 0.03–0.05 mg/kg	7–8	4
Medetomidine (i.m.)		
5 µg/kg	7–8	–
10 µg/kg	6–7	–
40 µg/kg	–	1–1.5
Alfentanil (i.v.)		
5 µg/kg	3–5	3–4
10 µg/kg	–	1–2

13.3 Premedicant and induction doses in the dog.

NSAIDs

Carprofen	4 mg/kg pre-operatively or intra-operatively. Lasts 18 hours
Ketoprofen	2 mg/kg s.c., i.m. or i.v. daily for up to 3 days or 1 mg/kg orally daily for up to 5 days
Meloxicam	0.2 mg/kg orally daily for up to 21 days, then 0.1 mg/kg
Flunixin	Useful in acute pain, and endotoxic or septic shock. 1 mg/kg, give by **slow** i.v. injection or s.c., but not i.m. Do not give more than 3 doses. Not suitable for pre- or intra-operative use
Ibuprofen	5–10 mg/kg orally, 24–48 hourly
Paracetamol	10–20 mg/kg orally given 6 hourly with codeine
Phenylbutazone	2–20 mg/kg per day orally in divided doses
Mefenamic acid	10–30 mg/kg post-operatively daily in divided doses
Piroxicam	0.3 mg/kg orally every 48 hours
Aspirin	For relief from milder pain. 25 mg/kg orally. Lasts 4–6 hours

CAT

The main consideration when anaesthetizing a cat is the need to avoid stress on induction, which can provoke laryngospasm. They should be fasted for 12 hours prior to induction to minimize the risk of vomiting.

Sedation/Premedication

1. *Medetomidine* 50–150 μg/kg i.m. produces sedation in 10–15 minutes and lasts 30–180 minutes depending on the dose used. It may provoke vomiting shortly after injection. It can be rapidly reversed with *atipamezole* at 2¹/₂ times the previous dose of medetomidine, i.e. if using Domitor and Antisedan, use half the volume of Antisedan to the volume of Domitor that was given.

2. *Xylazine* 1–2 mg/kg i.m. or s.c. gives good sedation for 30–40 minutes but provokes vomiting. It is useful in combination with ketamine for full anaesthesia (see below).

3. *Ketamine* 10–20 mg/kg i.m. gives moderate sedation with analgesia for 30–45 minutes but can be painful on injection. The increased muscle tone and lack of palpebral reflexes can be a problem, so the use of a bland ophthalmic ointment is important.

4. *Acetylpromazine* 0.05–0.1 mg/kg i.m. or s.c. (or 1–3 mg/kg orally 1 hour before sedation is required) is less reliable as a sedative but tranquillizes adequately prior to induction and reduces post-operative excitement due to barbiturates or alphaxalone/alphadolone. Note that it causes hypotension.

5. *Alphaxalone/alphadolone* (Saffan) 9–12 mg/kg i.m. produces light sedation. The large volume of injection needed for producing deep sedation or anaesthesia renders this a unsuitable method.

6. *Atropine* 0.05 mg/kg i.m. is advisable as a premedicant, especially if it is intended to intubate or to use ketamine.

General anaesthesia
Injectable agents.

1. *Alphaxalone/alphadolone* 9 mg/kg i.v. produces 10 minutes surgical anaesthesia. Thereafter, incremental doses of 3 mg/kg can be given or a continuous infusion at 0.2 mg/kg/min.

2. *Ketamine* at 20–30 mg/kg in combination with *xylazine* 1 mg/kg i.m. or s.c. will produce anaesthesia with good muscle relaxation. It is necessary to monitor the respiration and heart rates carefully.

   Ketamine 2.5–7.5 mg/kg i.m. can also be combined with *medetomidine* 80 μg/kg i.m. The onset of anaesthesia takes 3–4 minutes and lasts 30–60 minutes, depending on the dose of ketamine used. Atipamezole can be used to reverse the medetomidine (see above for dose).

3. *Barbiturates. Thiopentone* 10 mg/kg of a 1.25% solution given i.v. with care to ensure that it does not go perivascularly, or *methohexitone* 4–8 mg/kg of a 1 per cent solution i.v. will both give 5 minutes of light general anaesthesia. They are best given following acetylpromazine premedication and used for induction prior to inhalation anaesthesia.

4. *Propofol* at 7.5 mg/kg i.v. gives a few minutes anaesthesia with smooth rapid recovery. It is also useful for induction.

Inhalation anaesthesia. This is not a suitable method for *induction* of anaesthesia in the cat and modern injectable agents are much better, but it is a very satisfactory method of *maintaining* anaesthesia. To avoid laryngospasm on intubation, first spray the larynx with 2 per cent lignocaine then intubate with care using a 3–4 mm tube (for adults). T-piece or Bain circuits are ideal and suitable inhalation agents include halothane, methoxyflurane, isoflurane, and enflurane.

Long-term anaesthesia in the cat

A well-maintained inhalational technique with the animal intubated has much to commend it but if the experimental design does not permit this, injectable alternatives exist. Rapidly metabolized agents whose effects are non-cumulative are the best choice. Either: *alphaxalone/alphadolone* (Saffan) 9 mg/kg i.v. for induction followed by either incremental injections at 3 mg/kg or as a continuous infusion at 0.2 mg/kg min; or *Propofol* 7.5 mg/kg i.v. followed by incremental doses at 2 mg/kg thereafter.

Analgesia

The potential difficulties in providing safe analgesia for cats should not be used as an excuse to preclude the use of analgesics completely. Opioids can be used in cats but care must be taken not to over-dose. NSAIDs are generally poorly metabolized in cats and must also be used with caution.

Opioids.

Buprenorphine	0.01 mg/kg s.c. or i.m. lasts at least 8 hours
Nalbuphine	1.5–3.0 mg/kg i.v. lasts 3 hours
Pethidine	2.5–10 mg/kg s.c. or i.m. lasts up to 2 hours
Pentazocine	2 mg/kg s.c. or i.m. lasts at least 4 hours
Butorphanol	0.4 mg/kg s.c. lasts 3 hours
Morphine	0.1 mg/kg s.c. or i.m. lasts at least 4 hours

NSAIDs.

Ketoprofen	2 mg/kg s.c. daily for up to 3 days or 1 mg/kg orally daily for up to 5 days
Aspirin	10–20 mg/kg orally is best used only as a single dose as it shows some toxicity in the cat. It can be repeated if necessary but dose only every 72 hours maximum.
Phenylbutazone	1–5 mg/kg orally daily

Paracetamol is toxic in the cat.

FERRET

Ferrets should be fasted for 12 hours prior to induction to minimize the risk of vomiting.

Sedation

Medetomidine and atipamezole: 100 μg/kg medetomidine produces good sedation and loss of the righting reflex in 5–10 minutes. Domitor is reversed with an equal volume of *Antisedan* (1 mg/kg atipamezole) in a few minutes.

Injection anaesthesia

1. *Medetomidine* plus *ketamine.* 50–100 μg/kg medetomidine and 4–8 mg/kg ketamine given together i.m. produces stable surgical anaesthesia lasting 60 minutes or more. This is reversed by an equal volume of atipamezole i.m., e.g. for a 0.5–1 kg ferret, use 0.12 ml Domitor and 0.08 ml ketamine together, and reverse with 0.12 ml Antisedan.
2. *Ketamine* plus *xylazine.* 25 mg/kg ketamine i.m. and 1–2 mg/kg xylazine i.m. will give 30–60 minutes of good surgical anaesthesia.
3. *Saffan* (12 mg/kg i.v.) may be given into the cephalic vein if there is a competent person available to restrain the animal; or into the jugular vein after tranquillization with ketamine.

Other drugs

Atropine dose is 0.05 mg/kg i.m. to reduce salivary secretions.

Doxapram dose is 1–2 mg/kg i.v. to reverse respiratory depression. Repeat after 20 minutes if necessary.

Analgesia

Buprenorphine	0.05 mg/kg every 6–8 hours.
Ketoprofen	2 mg/kg daily

PRIMATES
Pre-anaesthetic medication

For smaller primates, such as marmosets, manual restraint may be sufficient to enable the i.p. or i.m. injection of anaesthetic directly. Premedication may only be required in order to facilitate intravenous injection or administration of inhalation agents.

Larger primates, such as the rhesus monkey, should be heavily sedated prior to handling or they can cause physical injury to the handler.

A crush cage is recommended for administration of sedatives.

1. *Ketamine* (5–25 mg/kg i. m.) is the drug of choice. Lower doses produce heavy sedation. Higher doses produce light anaesthesia. Peak effect is reached in 5–10 minutes and lasts 30–60 minutes.

2. *Alphaxalone/alphadolone* (12–18 mg/kg i.m.) is good for small primates, it produces heavy sedation. Additional doses i.v. produce surgical anaesthesia. Peak effect is reached 5 minutes after i.m. injection and lasts 45 minutes. For larger primates the volume that has to be injected is too large for it to be used.

3 *Acetylpromazine* (0.2 mg/kg i.m.) produces sedation but insufficient for safe handling.

4 *Diazepam* (1 mg/kg i.m.) also produces insufficient sedation for handling of larger primates.

5. *Fentanyl/fluanisone* (Hypnorm) (0.3 ml/kg i.m.) or fentanyl/droperidol (Thalamonal) (0.3 ml/kg i.m.) produces heavy sedation and good analgesia.

6. *Atropine* (0.05 mg/kg i.m.) should be given to reduce bradycardia produced by neuroleptanalgesia, and to reduce salivary secretions, especially when ketamine is used.

General anaesthesia

Primates should be fasted for 12–16 hours prior to general anaesthesia.

Injectable agents.

1. *Alphaxalone/alphadolone* (Saffan) (10–12 mg/kg i.v.) produces good surgical anaesthesia with fairly rapid recovery. Prolonged periods of anaesthesia can be provided by administration of additional doses every 10–15 minutes (5 mg/kg i.v.).

2. *Methohexitone* (10 mg/kg i.v.) or *thiopentone* (15–20 mg/kg i.v.) produce 5–10 minutes of surgical anaesthesia. However, analgesia is poor.

3. *Ketamine* (10 mg/kg i.m.) with *xylazine* (0.5 mg/kg i.m.) produces surgical anaesthesia with good muscle relaxation lasting 30–40 minutes. *Ketamine* (15 mg/kg i.m.) with *diazepam* (1 mg/kg i.m.) has similar effects.

Intravenous injection can be done in larger primates using the cephalic vein on the anterior forelimb or saphenous vein in the hindlimb. In marmosets, the lateral tail vein can be used. A butterfly needle is recommended.

NOTE. The dose of i.v. agents can be reduced by up to 50 per cent following premedication with ketamine.

Inhalational agents.

All common inhalational agents are suitable, e.g. *halothane*, *methoxyflurane*, *enflurane*, *isoflurane*. Usually, anaesthesia is induced by injectable agents, then inhaled agents used for maintenance. For small primates, a face mask may be sufficient; but for larger primates, intubation is recommended. Anaesthesia can be maintained using the volatile agent with oxygen alone, or with nitrous oxide in addition.

Intra-operative considerations

Body heat must be conserved during anaesthesia. The monkey should be anaesthetized on a heat pad or otherwise kept warm. Recovery should also take place in a warm environment. If anaesthesia is to last more than 1 hour, intravenous fluids should be administered, e.g. Hartmanns solution, or 0.18 per cent sodium chloride with 4 percent dextrose at a rate of 10 ml/kg. Fluids should be warmed prior to administration.

Analgesia
Opioids.

Buprenorphine	(0.005–0.01 mg/kg i.m.) is the agent of choice. Analgesia lasts 6–8 hours
Pentazocine	2–5 mg/kg i.m. lasts for 4 hours
Pethidine	(2–4 mg/kg i.m.) given immediately prior to recovery produces analgesia lasting 2–3 hours
Morphine	1–2 mg s.c. or i.m. lasts for 4 hours

NSAIDs

Ketoprofen	2mg/kg s.c. daily
Flunixin	2–10 mg/kg s.c. daily
Aspirin	20 mg/kg orally every 6 hours

PIG

When anaesthetizing the pig it is commonplace to tranquillize it first to facilitate handling. However, the need for this can be reduced by training of the animal. Pigs are highly intelligent animals and can readily be trained to accept a number of procedures. If they are not, the restraint will provoke much squealing which can be very stressful for everyone involved. Pigs usually have a thick covering of fat and care must be taken when giving i.m. injections that they are administered deeply enough. The usual site is in the muscles of the neck. Once asleep care must be taken to ensure that the airway is kept clear and laryngospasm is prevented. Applying pressure on the vertical ramus of the mandible, to push the jaw forward, and pulling the tongue forwards will help to prevent obstruction. Administration of atropine (0.3–2.4 mg depending on the size of the pig) to dry secretions will also help. Monitoring anaesthesia is facilitated by the pig's hairlessness, as the colour of the skin can readily be seen. However, this hairlessness also makes the pig susceptible to hypothermia in the post-operative period and steps must be taken in caring for the animal to prevent this.

Some strains of pig suffer from a biochemical myopathy known as porcine malignant hyperthermia. When anaesthesia is induced, a contracture of the muscles occurs, the pig goes stiff and its temperature starts to rise. If untreated the animal will die. The strains affected are Poland-China, Pietrain, and some lines of Landrace and Large White.

Tranquillization

Azaperone (Stresnil), a butyrophenone, is the drug of choice. The dose is 2–4 mg/kg by deep i.m. injection. The dose must be adapted according the the degree of excitation of the individual pig. The animal should be left undisturbed for 20 minutes following the injection or excitement may be provoked. It has no adverse effects on parturition, lactation, mothering instinct or food intake. It causes mild vasodilation and a slight fall in blood pressure. Other alternatives to consider are *diazepam* (2.0 mg/kg i.m.) or *acetylpromazine* (0.1 mg/kg i.m.), or a combination of *droperidol* (0.5 mg/kg) with *midazolam* (0.3 mg/kg). Inject the tranquillizer via a 21 gauge 1¹/₂ inch needle with a flexible extension line into the neck muscle to avoid the need for prolonged restraint. Following tranquillization, prior to intravenous administration of drugs to induce full anaesthesia, local anaesthetic cream (e.g. EMLA, Astra) can be applied over the site of venepuncture (ear) for 45 minutes to provide desensitization of the area. Less pain is caused if flexible cannulae are used in preference to rigid needles. Once the animal is quiet, an i.v. line can be put into the ear vein for induction with another agent, followed by intubation and maintenance with inhaled agents.

General anaesthesia

Apnoea is a common complication of surgical anaesthesia in the pig, especially if it is very fat. Its head should be kept at as natural an angle as possible to prevent pressure on the larynx, and endotracheal intubation should be considered to enable ventilation to be carried out if necessary. The use of local anaesthetic techniques in addition so that only a light plane of general anaesthesia is necessary will help to reduce this problem, which is seen when deeper levels of anaesthesia are reached.

The most satisfactory method of induction is the intravenous administration of an agent to an already tranquillized pig. There are many agents available, some of which can be used for maintenance of anaesthesia.

Metomidate (Hypnodil) i.v. at 3.3 mg/kg 20 minutes after aza-perone at 2 mg/kg i.m. will cause recumbency which lasts 15–20 minutes. There is poor analgesia and local anaesthetic agents may be necessary in addition, for some surgery to be carried out.

Alphaxalone/alphadolone (Saffan) at 6 mg/kg i.v. without pre-medication produces surgical anaesthesia for 10–15 minutes. If given following azaperone at 4 mg/kg i.m. a Saffan dose of 2 mg/kg i.v. is required. Incremental doses can be given for maintenance, there is minimal respiratory depression with good muscle relaxation and recovery is smooth.

Ketamine combinations. 2–5 mg/kg ketamine i.v. after 1 mg/kg xylazine i.m. produces good anaesthesia, as does 10-18 mg/kg ketamine i.m. after 1–2 mg/kg diazepam i.v.

Methohexitone at 5–6 mg/kg of a 2.5 per cent solution will produce anaesthesia in the unsedated pig with recovery in 10–15 minutes. Following premedication the dose should be reduced. Small incremental doses can be used to prolong the anaesthesia without affecting the recovery period.

Thiopentone i.v. at 5–10 mg/kg of a 2.5 per cent solution will induce anaesthesia. (Reduce the dose if a premedicant has been given.) Incremental doses will be cumulative and result in prolonged recovery.

Pentobarbitone at 30 mg/kg i.v. in the unsedated medium-sized pig will produce medium depth anaesthesia. For large pigs, the dose per kilo should be reduced.

Analgesics
Opioids.

Buprenorphine	0.05–0.01 mg/kg i.m. 8-12 hourly
Morphine	0.1 mg/kg i.m. 4 hourly up to a maximum of 20 mg
Pentazocine	2 mg/kg i.m. 4 hourly
Pethidine	2 mg/kg i.m. 4 hourly up to a maximum of 1g in large pigs

NSAIDs

Flunixin	1 mg/kg s.c. daily
Phenylbutazone	1g/225kg daily for four days, then on alternate days, orally.

SHEEP AND GOAT

The major consideration when anaesthetizing sheep and goats is that they are ruminants. It is important to prevent ruminal tympany by passing a stomach tube and leaving it in place throughout the operation. The degree of tympany will be reduced by keeping the animal only on hay for 24 hours pre-operatively. Regurgitation of rumen content is always a possibility, so endotracheal intubation with a cuffed tube is essential when anaesthesia is induced in adult ruminants. During procedures, keep the animal on its side or sternum. If kept in dorsal recumbency, pressure from the rumen will decrease venous return to the heart. Keep the head lowered so any saliva or rumen content can run out. Post-operatively, place the animal on its sternum and watch for signs of tympany.

The use of atropine is generally contraindicated in ruminants, as it renders salivary and bronchial secretions more viscous and harder to clear. If bradycardia develops during surgery, it can be given at 0.6–1.2 mg i.v.

To reduce contamination of the surgical field it is sensible to use sheep that have been shorn.

For **sedation** in combination with local or regional anaesthesia:

1. *Xylazine* 0.1–0.2 mg/kg (goats 0.05 mg/kg) i.m. gives 30–35 minutes of heavy sedation with analgesia.

2. *Acetylpromazine* 0.05–0.1 mg/kg i.m. produces sedation.

3. *Diazepam* 0.5 mg/kg i.m. or 0.25 mg/kg slowly i.v. produces good tranquillization.

General anaesthesia
Injectable agents

1. Injectable anaesthesia can be achieved with *ketamine* 10–15 mg/kg slowly i.v. but this produces high muscle tone and trembling. It is better to combine it with *xylazine* or *diazepam*. 0.2 mg/kg xylazine i.m. followed 15 minutes later by 10 mg/kg ketamine slowly i.v. will produce 40–45 minutes of surgical anaesthesia for sheep. For goats it is preferable to use a slightly lower dose of 0.1 mg/kg xylazine followed by 5 mg/kg ketamine slowly i.v. after 10 minutes to produce 15–20 minutes of anaesthesia. Diazepam 0.5 mg/kg i.v. followed by 4 mg/kg ketamine i.v. produces a short period of anaesthesia but is a useful method of induction causing minimal respiratory depression. Xylazine 0.05 mg/kg i.v. or 0.1 mg/kg i.m. followed by ketamine 5 mg/kg i.v. is also a good method for induction prior to intubation and inhalation anaesthesia.

2. *Thiopentone* 10–15 mg/kg i.v. will last 5–10 minutes to allow intubation.

3. *Alphaxalone/alphadolone* (Saffan) 2.2 mg/kg i.v. followed by incremental doses of 1–2 mg/kg i.v. every 15–20 minutes will produce stable long-term anaesthesia.

4. *Propofol* 4–6 mg/kg i.v. in sheep or 3–4 mg/kg i.v. in goats

is useful for induction. Incremental doses can be given for maintenance.

5. *Pentobarbitone* can be used i.v. for induction at 20 mg/kg prior to inhalation anaesthesia. Incremental pentobarbitone at 2 mg/kg every 5 minutes can be used to maintain anaesthesia but recovery will be prolonged and there will be marked respiratory depression necessitating oxygen supplementation. Some commercially prepared solutions of pentobarbitone contain propylene glycol which has been reported to cause haemolysis in sheep and goats.

6. *Methohexitone* at 4 mg/kg i.v. can be used for induction and incremental doses of 50–75 mg per minute used for maintenance. However, there is respiratory depression as with the other barbiturates and there may be twitching and convulsive activity on recovery unless sedative premedicants have been given.

Inhalational anaesthesia. Following induction with an injectable agent, the sheep should be intubated to prevent any aspiration of saliva or ruminal content. To intubate, spray the vocal cords with lignocaine and use a suitable laryngoscope blade. Maintain on a Magill or Bain circuit. For combinations of nitrous oxide and oxygen, flow rates of 6l/min and 2l/min respectively will be suitable.

Analgesics
Opioids.
Buprenorphine	0.01 mg/kg lasts 4–6 hours
Butorphanol	0.2 mg/kg lasts several hours
Morphine	10 mg total dose s.c. or i.m. lasts 2–3 hours
Pethidine	200 mg total dose i.m. lasts 2–3 hours

NSAIDs.
Flunixin	2 mg/kg s.c. daily

These α_2-**agonists** also provide analgesia but do not last very long:

Xylazine	50–100 μg/kg
Detomidine	10–20 μg/kg
Clonidine	6 μg/kg

Regional blocks with local analgesics, such as lignocaine, are very useful in sheep and goats.

Special considerations for lambs and kids
As with any neonatal animal the time away from the mother should be kept to the minimum possible and care taken to prevent the

animal from becoming hypothermic. Sedatives are better replaced with short-acting agents, so the animal returns to normality as soon as possible. The dose of xylazine should be reduced to half the adult dose for lambs and to 25 μg/kg i.m. for kids. The use of atipamezole to reverse the effect of xylazine will help to speed recovery. Avoid the use of barbiturates, especially pentobarbitone, since they are not metabolized as rapidly as in the adult.

BIRDS

Several aspects of their behaviour, anatomy, and physiology require consideration when giving an anaesthetic to a bird. They have a very high basal metabolic rate compared to mammals and a higher body temperature (40–44°C). This means there is a high rate of food conversion resulting in rapid onset of hypoglycemia when a small bird is deprived of food. The rapid heart rate is subject to profound alteration in response to stress. Additionally, the difference between ambient temperature and body temperature will be great, and thermoregulation is less efficient and less adaptable in birds than in mammals.

Prevention of heat loss is therefore vitally important during avian anaesthesia. A heat source should be provided, insulation provided (such as bedding or wrapping in aluminium foil), a minimum of plumage removed (with due regard for maintaining sterility), and as little water or alcohol-based products as possible used in skin preparation to prevent cooling. In-flowing gases should be warmed and the rectal temperature of the bird monitored.

Acute cardiovascular failure may occur, especially in small birds, from postural hypotension if the bird is maintained on its back for long periods or if sudden alterations in posture are made. Damage to the brachial or lumbo-sacral plexuses due to over extension of wings or legs, or due to struggling will also cause cardiovascular failure. All movements of the bird must be made slowly and gently.

Injectable anaesthesia
Injection sites.
Intraperitoneal. These are given in the midline, halfway between the cloaca and the sternum. A 25 gauge needle is inserted at a shallow angle directed cranially so the point lies paralled to the abdominal wall. Care must be taken to avoid the air sacs.

Intramuscular. The pectoral muscles, or the thigh in a larger bird, are used. Care is needed not to hit a blood vessel.

Intravenous. For larger birds, the brachial veins are used (see Chapter 10). The walls are quite fragile.

If possible, weigh the bird prior to injection, but without increasing the stress by excessive handling. For example, placing it in a suitable narrow bag will aid restraint for this purpose.

Ketamine combinations. For poultry or wildfowl, 15–50 mg/kg will give light surgical anaesthesia. It may be given in incremental doses of 5 mg/kg at 5 minute intervals and can be used to maintain light anaesthesia/sedation for several hours.

For small birds, mix 0.1 ml ketamine (100 mg/ml) in 0.9 ml sterile water. Give 1 mg (0.1 ml)/bird i.m. This will produce sedation in 3–4 minutes and recovery in about 20 minutes. At 2 mg/bird, there is light surgical anaesthesia for 5–12 minutes and recovery in 30 minutes. At 3 mg/bird, there is 5–20 minutes of anaesthesia and recovery in 60 minutes.

Ketamine may be used alone for induction prior to using volatile agents for longer procedures, but is most useful when used in combination with other drugs. For deeper anaesthesia for surgical procedures where some relaxation of the muscles is required, ketamine at 20 mg/kg i.m. may be combined with midazolam (4 mg/kg i.m.), or diazepam (1.5 mg/kg i.m.), or xylazine (40 mg/kg), or acetylpromazine (0.5 mg/kg i.m.). The combination with midazolam provides the best level of analgesia. With xylazine there is marked respiratory depression and bradycardia and on some occasions birds have died up to 48 hours later, especially if this mixture is used together with methoxyflurane inhalation anaesthesia.

Saffan. This may be used i.v. in larger birds. 10–14 mg/kg will give about 5–10 minutes of surgical anaesthesia and further incremental doses may be given.

Barbiturates. These agents are not satisfactory for avian anaesthesia because of their narrow safety margins.

Propofol. Due to their high basal metabolic rate, this drug only lasts for a very short time, and may not provide time for intubation.

Other drugs. Atropine should be administered to decrease secretions which may block the airways. The dose is 0.05 mg/kg i.m. For respiratory depression use doxapram 5 mg/kg i.v.

Inhalation anaesthesia
If halothane is used for induction, extreme care must be taken since the induction is very rapid (awake→asleep→dead in a small bird may take just 45 seconds). Methoxyflurane will produce slower induction and is therefore much safer. It also produces better analgesia than halothane, but it is 50 per cent metabolized by the liver resulting in a prolonged recovery time. The inhalational anaesthetic of choice for

birds is isoflurane, as only 0.3 per cent is metabolized so recovery is more rapid. It is a good analgesic so a lower plane of anaesthesia is necessary than with halothane, with improved muscle relaxation and less respiratory depression. Following induction, the bird may be intubated and maintained on O_2/N_2O/halothane or isoflurane preferably via a T-piece system. Gas flow rates should be high, about three times the minute volume. For example:

Chicken   Bodyweight 2.5 kg = 770 ml minute volume

Pigeon   Bodyweight 300 g = 250 ml minute volume

Cage bird   Bodyweight 30 g = 25 ml minute volume

The airway is easily obstructed and care must be taken to keep it clear. Even short periods of apnoea can result in severe hypoxia, especially in small birds. A small plastic tube placed in the oesophagus will draw up fluid by capillary action and prevent it being aspirated into the glottis. Gases may be given directly into the interclavicular air sac by catheterization, but the trachea will be unguarded and aspiration may occur.

Following gaseous anaesthesia, oxygen must be administered to flush anaesthetic from the air sacs as it is released from the circulation. If the bird is not properly ventilated and this is not cleared, it will be re-absorbed and could result in fatal overdose.

Post-operative care

For small birds a quiet, dark, recovery box should be maintained at 40°C, and the bird adequately monitored. The wings should be fixed (taped to the back with micropore tape or the bird wrapped up in a towel and then taped) to prevent damage from flapping during the excitatory part of the recovery phase. Fluids should be administered at the rate of 5 ml/kg/hour. Isotonic or dextrose saline should be given s.c. to small birds or i.v. to larger ones. Haemorrhage may be significant in a small bird whose blood volume is 100 ml/kg. A 20 g bird therefore has a blood volume of 2 ml. Losing 5 drops of blood will represent a 15 per cent blood loss which will decrease venous return causing hypotension and cardiac arrest. The bird should be encouraged to eat as soon as possible and isoflurane induced/maintained birds will eat more rapidly than others, who may be left very depressed post operatively. Milupa baby food is a useful food to give to convalescent birds. Perches should be removed or lowered for birds in the recovery phase to avoid injury.

Analgesics

Butorphanol   3–4 mg/kg is the opioid analgesic of choice

Ketoprofen   2 mg/kg daily

Flunixin   1–5 mg/kg

Morphine	15–30 mg/kg
Codeine	2.5–30 mg/kg

Xylocaine may be used for local analgesia.

ANAESTHETICS/ANALGESICS AND RELATED DRUGS

Drug name	Trade name	Manufacturers
Acetylpromazine (Acepromazine)	ACP	C Vet Ltd
Alfentanil	Rapifen	Janssen
Alphaxalone/ alphadolone	Saffan	Pitman Moore
Atipamezole	Antisedan	SmithKline Beecham
Atropine	Atropine injection	C Vet Ltd
Azaperone	Stresnil	Janssen
Bupivicaine	Marcain	Astra
Buprenorphine	Temgesic	Reckitt & Colman
Butorphanol	Torbugesic, Torbutrol	Willows Francis
Carprofen	Zenecarp	C Vet Ltd
Chloral hydrate + pentobarbitone + magnesium sulphate	Equithesin	
Codeine/paracetamol	Pardale-V	Arnolds
Detomidine	Domosedan	SmithKline Beecham
Diazepam	Valium	Roche
Diclofenac	Voltarol	Ciba-Geigy
Doxapram	Dopram V	Willows Francis
Droperidol	Droleptan	Janssen
Enflurane	Ethrane	Abbott
Etomidate	Hypnomidate	Janssen
Etorphine/ acepromazine	Immobilon (large animal)	C Vet Ltd
Etorphine/ methotrimeprazine	Immobilon (small animal)	C Vet Ltd
Fentanyl	Sublimaze	Janssen
Fentanyl/droperidol	Thalamonal	Janssen
Fentanyl/fluanisone	Hypnorm	Janssen
Flunixin	Finadyne	Schering Plough
Halothane	Fluothane	Pitman Moore
	Halothane	Rhone Merieux
Ibuprofen	Brufen	Boots
Isoflurane	Forane	Abbott
Ketamine	Vetalar	Parke Davis
	Ketaset	Willows Francis

Ketoprofen	Ketofen	Rhone Merieux
Lignocaine	Xylocaine	Astra
	Lignavet	C Vet Ltd
Medetomidine	Domitor	SmithKline Beecham
Mefenamic Acid	Ponstan	Parke Davis
Meloxicam	Metacam	Boehringer Ingelheim
Methohexitone	Brietal	Elanco
Methoxyflurane	Metofane	C Vet Ltd
Metomidate	Hypnodil	Janssen
Midazolam	Hypnovel	Roche
Nalbuphine	Nubain	DuPont Pharmaceuticals
Naloxone	Narcan	DuPont Pharmaceuticals
Paracetamol/codeine	Pardale-V	Arnolds
Pentazocine	Fortral	Sanofi Winthrop
Pentobarbitone 60mg/ml	Sagatal	Rhone Merieux
Pentobarbitone 200mg/ml (for euthanasia)	Euthatal	Rhone Merieux
	Euthesate	Willows Francis
	Expiral	Sanofi Winthrop
	Lethobarb	Duphar
	Pentoject	Animal Care
	Pentobarbitone injection	Loveridge
Pethidine	Pethidine	Martindale
Phenylbutazone	Equipalazone	Arnolds
	Phenogel	Duphar
Piroxicam	Feldene	Pfizer
Propofol	Rapinovet	Pitman Moore
	Diprivan	Zeneca
Thiopentone	Thiovet	C Vet Ltd
	Intraval	Rhone Merieux
Xylazine	Rompun	Bayer
	AnaSed	C Vet Ltd

14 Surgical techniques

Before starting to carry out any surgical procedure, think carefully about the facilities that are available and whether they are adequate for what you wish to achieve. Ensure you are completely familiar with the anatomy of the species and the part that you will be working on. Practise by dissecting cadavers before obtaining your licence and attempting the procedure on a live animal.

PREPARATION FOR SURGERY

Prior to 1860, surgery was carried out with almost no preparation. The instruments were not washed, the patient's skin was not cleaned, and doctors did not wash their hands or their equipment. The hallmark of a 'good' doctor was a coat which was so caked in blood and tissue fluid that it would stand up on its own. The major problem associated with surgery under those conditions was sepsis, and patients frequently died from infection.

Then in 1860, Lister determined that there was a link between contamination of wounds and infection, and set about changing the way things were done. He advocated two things:

1. Removal of necrotic (dead) tissue and dirt from wounds.

2. Application of chemical disinfectants, such as carbonic acid, to the surgeons hands and to the wounds.

This was the start of **antiseptic surgery**.

Over several decades, antiseptic surgery developed into **aseptic surgery**. Here, instead of trying to remove infection from wounds, the idea was to prevent its occurrence by ensuring good sterility of the operating room, the equipment, the surgeon, and the patient. In the late 19th century, surgeons began to wear clean gowns, caps and gloves, and in the early 20th century masks were worn and doctors began to sterilize their instruments. This was when modern surgery really began.

PRINCIPLES OF ASEPSIS

The idea behind asepsis and aseptic technique is to remove all possible sources of contamination or infection from the surgical field prior to surgery, so as not to contaminate the wound. The main sources of contamination are:

- the atmosphere
- the surgical team
- the instruments and equipment
- the skin of the patient

PREPARATION OF A ROOM FOR USE AS AN OPERATING THEATRE

The atmosphere of the operating room is the source of contamination which is most difficult to control. Dust and other airborne particles can settle in the wound, bringing bacteria with them. In order to reduce this problem, there must be less dust and less air movement. Operating theatres should therefore contain as few fixtures and fittings as possible, to reduce the number of places where dust can collect, and they must be thoroughly cleaned and disinfected prior to surgery. This is facilitated in a well-designed theatre in which there are few ledges or cupboards, and where all fittings are flush with the walls so there are no small areas for dust to collect. Even non-recovery procedures depend on good surgery which requires precision and clarity of thought and all unnecessary clutter should be kept away. The best ventilation system is one which uses an input fan with a filter. This produces positive pressure in the room, and tends to push dust particles out of the room away from the patient. There must be adequate space around the operating area for all the equipment and instruments which you intend to use. The room should be maintained as a clean, dust free area. Dirt is brought in and is stirred up when people enter and leave, so as few people as possible should be allowed into the room. The operating room must not be used as a store room or a thoroughfare. The best way to maintain the operating room is to keep it as just that—an operating room and nothing else. Ideally, major surgery should be performed within a suite of rooms, each room having its own function:

1. The 'prep room' is the area where animals should be prepared for surgery. They are brought to this room from the animal room, kennel, pen or paddock, anaesthetized, clipped, and cleaned. Only once they are ready, do they go through into the theatre.

2. The *operating theatre* is used just for surgery. Care must

be taken not to bring dust, dirt, and hair through from the prep room.

3. *A recovery area* should be provided where animals can be taken after surgery for intensive nursing care and observation while recovering from anaesthetic. This room needs to be easily accessible for people, unlike the operating room.

The whole suite of rooms should be made with impervious floors, walls, and ceilings to allow thorough cleaning and fumigation if required.

It is also essential for the operating room to be equipped with a good, shadowless light source, to enable the surgeon to see clearly what he is doing, a steady table with facilities to adjust the height and tilt, a chair which can also be adjusted, and for surgery on very small animals or for microsurgery, a binocular microscope. An anglepoise lamp is not satisfactory since it can produce excessive heat over a long period of time, and casts shadows which make working difficult. For large animal surgery there must be adequate equipment for lifting and moving an anaesthetized or recumbent animal, without injury to it or to the handler.

PREPARATION OF THE SURGICAL TEAM

The bodies of the surgical team are potential sources of contamination of the wound. The resident population of bacteria and other microbes which are found on the skin may be pathogenic if they penetrate into a wound, and the surgeon can spread infection from one animal to another on his hands.

To prevent infection from being spread by the surgical team, the day's operations should be **planned in advance** so as to proceed in a logical order. Dirty (non-sterile) operations and autopsies should not be done before clean (sterile) surgery: it makes better sense to do clean operations first then move on to the dirty ones. This reduces the risk of spread of infection between animals. To control contamination from the skin of the surgeon, his or her body should be encased in a 'sterile cocoon'. This is achieved by **covering the entire body in sterile material**.

First, the surgeon and all other people intending to enter the operating theatre should arrive clean, with washed hands and not wearing jewellery. It is usual for all clothing, including footwear, to be either changed for 'scrub suits' used only in the theatre, or completely covered up. A cap should be worn which covers all hair, and a face mask must be worn. Masks protect the animal from pathogens in the respiratory tract of the people in the room, and also protect the people from pathogens coming from the animal. Once everyone in the operating room has changed into the appropriate clothing, the surgeon and those involved directly in the operation encase themselves in their sterile cocoons.

For full aseptic technique, the following protocol should be used. For minor surgical procedures, less stringent precautions need be taken and the protocol can be modified accordingly.

1. **Scrubbing up.** A 5 minute wash of hands and forearms removes many microbes from the skin. First, the hands and forearms, including the elbows, should be washed using antibacterial skin cleanser and running water, and the fingernails should be scrubbed. The hands should always be held above the elbows to prevent dirty water from running down the arm on to clean areas.

   Then the wash is repeated. The antibacterial skin cleanser should be massaged well into hands and forearms, and worked down to the elbows. Having done the elbows, the hands and arms should be rinsed from hand to elbow.

   The next stage is the 1 minute scrub. A sterile scrubbing brush is used to scrub the fingernails, fingers, and palms with antibacterial skin cleanser for 1 minute. Then the cleanser is massaged down the forearms, omitting the elbows.

   After rinsing, the final stage is to wash the hands and forearms again with antibacterial skin cleanser, without touching the elbows, then finally rinse with running water from hands to elbows. Hands may or may not be dried with sterile towels.

   After scrubbing up, the hands should always be held together in front of the body above the elbows, to prevent them from touching any non-sterile material and to prevent contamination running down the arm from the elbow.

2. **Gowning.** If sterile gowns are worn, they are autoclaved prior to use, and usually come in bags packed inside out. The bag should be opened by an assistant without touching the contents, and the surgeon takes the gown by the collar and pulls it out if the bag. The gown is held up and shaken lightly; this should allow the surgeon to see the armholes. The surgeon puts the gown on without touching the outside of it, and the ties at the back are done up by an assistant.

   Once the surgeon has his gown on, he is in his sterile cocoon and must not be touched by non-sterile objects or personnel.

3. **Gloving.** Surgical gloves come sterile in packs which are easily opened by an assistant without compromising sterility. They are packed already dusted with talc to facilitate putting them on, and a sachet of extra talc may also be provided. They come with the cuffs turned back, and the

surgeon picks them up from the bottom of the cuff so as not to touch the outside of the glove. For lengthy procedures, surgeons sometimes wear two pairs of gloves.

PREPARATION OF INSTRUMENTS AND EQUIPMENT

Sterilization or disinfection of instruments and equipment is mandatory prior to surgery, and there are many simple ways to achieve sterility.

Sterilization is the removal or destruction of all living microbes, including spores.

Disinfection is the removal or destruction of all living microbes, but excluding spores.

All disinfectants and sterilizing agents act by denaturing proteins.

Drapes and swabs can be bought sterile in packs. They are more expensive this way but much more convenient and remove the necessity for an autoclave. After use, instruments must always be washed thoroughly. Cold water will facilitate the removal of blood, and care must be taken to clean inside the joints and teeth of instruments where blood and tissue fluids collect.

Before sterilization, instruments should be washed in hot water, with detergent, to remove grease. Grease acts as a barrier to the penetration of chemical and physical sterilizing agents, and so must be removed. Again, particular care must be taken at the joints, where grease collects.

After sterilization, instruments should be allowed to dry, to prevent rusting, for example in hot air.

Physical methods of sterilization
Heat is the main physical method, either wet or dry.

1. *Boiling water.* Instruments which are boiled at 100 °C for 20 minutes will be free of most pathogens, but not of spores. This method is not suitable for fabrics, such as swabs, dressings, and drapes. Care must be taken that the instruments are allowed to cool before use or the skin of the animal may be damaged.

2. *Autoclaves.* These utilize steam under pressure. This method kills spores and sterilizes if carried out correctly. The time, pressure, and temperature used varies with the material. Instruments are usually done at 121 °C (15 pound-force per square inch) for 20 minutes. Autoclaves producing higher temperatures and pressures take less time. Rubber and some plastics can be autoclaved at lower temperatures, but take longer. The steam used in autoclaves penetrates through

fabrics, so this method is useful for drapes, gowns, swabs, and dressings.

Items to be autoclaved may be free in the autoclave, or prepacked into autoclave bags made of plastic, paper or cloth. Such bags allow the penetration of steam but not microbes, and can be used for storage of items after autoclaving. Several rules must be followed when using the autoclave:

(a) Steam must first displace the air before the sterilization cycle can begin. Some machines use a vacuum to remove the air.
(b) Steam must be allowed to penetrate through to the items inside the machine. Wrappings must therefore be of suitable permeable material, specifically designed for use in autoclaves.
(c) Fabrics must be folded loosely and not tightly packed or steam will not penetrate. If the autoclave is too full, it will not run efficiently.
(d) Indicators placed within the packaging should be used to check that sterility has been achieved.
(e) Steam must be allowed to escape from the packaging after autoclaving so the contents may dry out.

Once they leave the autoclave, packets containing instruments have only a limited shelf-life during which they remain sterile. The date of autoclaving should therefore be written on each packet and old packets re-sterilized at intervals.

Autoclaving is not suitable for catheters, endoscopes, and delicate materials.

3. *Dry heat.* Hot-air ovens held at 160 °C for at least one hour will sterilize items. This is only suitable for some metal and glass articles. The method is efficient at removing pathogens, but slow. It is useful because it does not cause corrosion of metals or pitting of glass which can occur with repeated autoclaving. Also, instruments can be sterilized coated with grease, as the grease does not prevent the object from being heated. It is used for sharp optical instruments, as it does not cause blunting. It is not suitable for fabrics, as these simply burn in ovens at this temperature.

4. *Burning.* This is used for destruction of contaminated materials.

5. *Ionizing radiation.* This is used commercially for sterilization of needles, syringes, suture materials, and delicate

items. The method is very efficient, but not suitable for use in a standard operating theatre.

Chemical sterilization

Chemical methods are less efficient than physical means, but are often used for delicate items. Instructions must be followed carefully for good results. Several factors must be taken into consideration before using chemical sterilizing agents:

(a) Items to be sterilized must be very clean, or the chemical will not penetrate. Many chemicals are inactivated by debris, such as blood.

(b) The method is slow, and sufficient time must be allowed for sterilization to occur.

(c) Liquid sterilizers must be at the correct concentration. A higher concentration does not necessarily mean more efficient sterilization.

(d) Most chemical agents are irritant, so all residues must be removed prior to use of the instrument.

(e) For gaseous agents, items must be dry and moisture-free.

Chemical agents are often used as quick sterilizing agents in field situations, where there is no access to an autoclave.

Chemicals used for sterilization are as follows:

1. *Ethanol.* Ethanol at 70 per cent is disinfectant. Higher or lower concentrations are less effective.

2. *Phenolics.* The cold sterilizer Novasapa (Willows Francis Veterinary) contains chlorocresol which is a phenolic disinfectant; triethanolamine, a cleansing agent; sodium formate, a descaler; and sodium citrate to prevent blood from clotting on the instruments. If instruments are immersed in it for 1 minute they will be free from gross contamination and if placed in boiling Novasapa, spores will also be killed. Aqueous or organic iodine preparations are disinfectant. They corrode metals but are useful in emergencies, e.g. Pevidine.

3. *Quaternary ammonium compounds.* Aqueous or organic solutions are used. They are relatively non-toxic and non-irritant, e.g. chlorhexidine (Hibitane, Savlon), benzylkonium chloride (Marinol Blue, BK Vet). Hibitane in methylated spirit forms quick-sterilizing fluid, and items are placed in this for 3–4 minutes.

4. *Glutaraldehyde (Cidex).* This disinfects in 3–4 minutes, and sterilizes in 12–24 hours. It is useful for optical instruments.

5. *Formaldehyde gas.* This is slow, taking 24 hours. It is unpleasant to use, and items must be rinsed in sterile water after use. Instruments must be free from organic material before use or this will be fixed on to the instrument by the formaldehyde.

6. *Ethylene dioxide.* This is used for large, bulky pieces of equipment. It is suitable for items which cannot be autoclaved, and many pre-packed items have been sterilized in this way. Items must be aired prior to use to allow the ethylene dioxide to escape.

PREPARATION OF THE ANIMAL

The body of the animal carries two types of contaminants:

1. Resident commensal or symbiotic flora, which are hard to eliminate.

2. Extraneous potential pathogens, which are easier to eliminate.

These bacteria feed on debris on the surface of the skin. Prior to surgery, therefore, a form of mechanical cleansing is required to remove this debris.

Animals covered with fur need to be clipped to allow cleansing of the skin. Ideally this should be done 24 hours before surgery, to allow the skin to recover from any trauma and to allow loose hair to fall off. If shaving is done immediately prior to an operation, care must be taken to remove any loose hair. This is most satisfactorily done using a small vacuum cleaner. Clipper blades must be clean and sharp to minimize any damage to the skin, and an adequate area must be shaved to prevent contamination of the wound.

The animal must be appropriately positioned for surgery. Ensure that there is no restriction to respiration by extending the head and neck, and position the limbs such that they do not cross the chest. There is a variety of commercial cradles available for positioning animals so that the use of limb ties is rarely necessary. If these are used, great care must be taken to ensure that they are not applied too tightly. Inexpensive animal positioning devices can be made from pieces of foam rubber or sand bags which can be covered in washable material and they will also help to provide some insulation around the animal, reducing heat loss. If the procedure is to be prolonged, bony areas should be padded to prevent pressure sores, bland ophthalmic ointment put into the eye to prevent corneal drying, and in ruminants a stomach tube passed to prevent ruminal tympany.

There are three stages in the preparation of the skin:

1. A wash to remove any gross debris, such as mud or faeces. Be sure to use a cleaning agent that is compatible with the antiseptic agent you will be using.

2. An antiseptic wash, with a quaternary ammonium compound (Savlon or Hibitane), or an organic iodine preparation (Pevidine). These must be used at the recommended concentrations to avoid causing any tissue damage.

   It is particularly important with small laboratory animals to avoid hypothermia, so solutions used for preparation of the skin must be warm, and the fur must not be allowed to get overwet. Skin should be cleansed from the middle of the clipped area towards the outside, never the other way round, or this will drag dirt from the fur at the edge on to the skin where the incision will be.

3. Alcohol may be applied to the skin as a final preparation. Again, in small animals this may be omitted to prevent excess heat loss.

After the skin has been cleaned, sterile drapes are applied, by the surgeon, to cover all but the surgical field. Fabric or paper drapes may be used. Drapes must be dry, or they act as a wick drawing contamination up from beneath and on to the surgical site.

A final consideration in maintaining asepsis is whether to use prophylactic antibiotics. In theory, these should not be necessary if aseptic technique is good, but in practice they are often used. Efficacy is enhanced if these are given pre- or peri-operatively, rather than post-operatively, so that the drug is already working when the contamination occurs.

AVOIDANCE OF POST-OPERATIVE INFECTIONS

Factors which favour the development of post-operative infections are as follows:

1. *A high level of contamination*, from dirty equipment, a dirty room, dirty hands, and poorly cleaned wounds.

2. *Injured tissues*. Poor surgical technique results in excessive bruising. Damaged tissues cannot fight infection well.

3. *Bleeding*. Blood is a perfect medium for the growth of bacteria.

4. *'Dead space'*. Pockets will be left between tissues after surgery (if wound closure is inadequate) where blood and serum can accumulate predisposing to infection.

5. *Stress*. This reduces the efficiency of the immune system.

6. *Immunosuppressant drugs*, e.g. corticosteroids.

7. *Some inbred strains* have reduced immunocompetence.

There is a common misconception by some research workers that rats are in some way resistant to post-operative wound infections and therefore much experimental rat surgery is performed under non-sterile conditions. Bradfield *et al.* (1992) have shown that, although the rat may not show clinical signs of infection, those with high wound bacterial counts show changes in behavioural, biochemical, and histological parameters which may confound experimental data. In most cases there is room for improvement in the sterile technique of rodent experimental surgery, for the benefit of the science, even if not for the benefit of the animal.

SURGICAL TECHNIQUES

A surgical procedure should be as atraumatic as possible. It is important to learn how to handle instruments correctly and to avoid damaging tissues by grasping them with excessive force or with inappropriate instruments. Cutting instruments must be kept sharp and crushing instruments must not be used on tissues which are to be left *in situ*. For examples of some common instruments see Chapter 16. There are many different companies producing surgical instruments (see Directory). Looking through their catalogues will help in the selection of appropriate instruments for the procedure you wish to carry out. Instruments must also be a suitable size for the species and area on which you are working.

Usually the skin incision will be made using a sharp (i.e. new) scalpel blade. The scalpel should be held like a pen and the tissue cut with a firm single stroke. Subcutaneous tissues can then be separated and cut with scissors, using the points to spread tissue and separate it into its distinct layers and then cutting them with the blades. While dissecting the tissue it is best to hold it with rat-toothed forceps or more delicate tissue may be held with blunt dissecting forceps. Some tissues, such as a fatty liver, cannot be satisfactorily held with forceps at all because of the risk of damage and haemorrhage. The most common haemostatic forceps used to grasp bleeding vessels are Spencer–Wells artery forceps or Kocher's forceps which combine crushing jaws with pointed tips, and are particularly useful for picking up bleeding vessels which have retracted into tissue, such as fat.

If you are operating to remove tissue, and do not need to preserve it for histology or other examinations, it is possible to grasp it with larger and more powerful instruments. In general, try to hold the tissue which is to be discarded, and touch the tissue which is to be left *in situ* as little as possible. In order to avoid bruising, tearing, and stretching of tissue, it is preferable to make a larger, rather than a smaller incision. Wounds heal from side to side, not from end to end and therefore a larger incision which allows adequate vision is always better and will not have any effect on the rate of healing. An incision made with sharp instruments with minimal handling of tissue and sutured appropriately will heal rapidly with minimum discomfort to the animal.

HAEMOSTASIS

Haemostasis can be achieved by picking up bleeding vessels and tying them off with a simple ligature using as fine a material as possible (but providing adequate strength to allow tying a secure knot) or by the use of diathermy, radiosurgery or cautery, all of which apply heat to the vessels to coagulate them. Large vessels require ligatures for secure haemostasis although it may be possible to seal them effectively enough with diathermy to prevent retrograde haemorrhage from tissue which is about to be discarded. Metal clips can also be used for larger vessels but they are expensive and require special instruments. Ligatures for haemostasis should be made of absorbable materials such as catgut or Vicryl (see Chapter 15). The use of non-absorbable materials, such as nylon, can act as a focus for infection, which may lead to the formation of a discharging sinus or give rise to peritonitis if used in the abdomen.

Before starting any surgical procedure involving entry to a body cavity it is important to count the number of swabs provided and the number of instruments, paying particular attention to the number of any small items such as loose suture needles. Any additional swabs or other items provided during the procedure should also be noted in writing by an assistant and before the wound is closed it is important to check carefully that everything has been accounted for.

The amount of blood loss should be carefully monitored, as a minute volume is potentially very significant in a small animal. In larger animals it is useful to weigh the swabs used to give an estimate of blood loss so fluids can be infused at an appropriate rate.

Whatever the size of the animal, good control of haemorrhage is always important in order to avoid post-operative swelling and to minimize surgical shock and risk to the animal. Haemorrhage will be associated with pain and poor wound healing. To assist control of bleeding, instruments need to be in good condition with jaws that meet properly. To check the jaws, hold the instrument up to the light and look between the closed jaws for distortion or gaps. Good quality instruments are always worth the additional cost, and they must be looked after with care to ensure that blades are not damaged. Surgical instruments must be used only for surgery.

Surgical procedures should always be kept as short as possible and care must be taken to stop tissues from drying out under the heat of the operating lights and from exposure to the air. Post-operative infection is often related to the length of the procedure. Tissues should be protected from drying by covering them with sterile gauze swabs soaked in warmed saline or by regular applications of warmed sterile saline directly on to the exposed viscera.

Surgical procedures should be scheduled for a time in the day and a time in the week when there will be adequate staff available to provide good post-operative care.

15 Suturing techniques and materials

When closing a wound, it is necessary to insert sutures in order to restore the anatomical structure of the tissues and support them during the healing process. Sutures are generally placed using sterile needles, needle holders, and tissue forceps. Needle holders come in many patterns and sizes (see Figure 16.8) and the choice will depend largely on personal preference. Whichever one is chosen, it is essential to know how to hold it properly and how to manipulate it with dexterity. Practice is essential for good suturing technique. Inefficient use of the needle holders will lead to wasted time and poor suturing. The licensee should observe those who are skilled, and the technique should be practised. Freshly incised banana skins or foam rubber make good models for practice.

SUTURE PATTERNS

The general aim when suturing is merely to appose the tissue margins to allow healing—excessive tension on sutures should

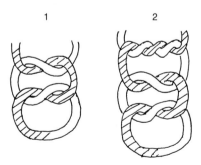

15.1 Knot used with coated Vicryl. A simple square (reef) knot (1) will allow the knot to be glided down into place and will hold until a third throw (2) is placed to lock the knot. Each throw should be snugged down with gentle pressure until it is tight.

be avoided as it will slow healing. There are many specialized suture techniques for particular tissues and specialized surgical texts should be consulted for these.

The basic *surgical knot* is common to all suture patterns. This can be hand- or instrument-tied and it is helpful to become familiar with both methods. Essentially it is a square (reef) knot with an extra throw, i.e. right over left, left over right, right over left. Three throws are sufficient for most suture materials but for some slippery suture materials, or in wounds where there is some tension, it helps to use a surgeon's knot, i.e. the first throw is a double one. Some suture materials are better if used with a modification of this knot (for example, see Figure 15.1 for Vicryl and Figure 15.2 for PDS—polydioxanone).

Knot security is obviously important and depends on the type of knot, the coefficient of friction of the suture material, and the technique used for tying the knot. Some sutures are coated with wax or silicone to reduce tissue drag. This will reduce friction and decrease knot security. Thus, for Vicryl and PDS, special knots are used. The simplest knot that will provide security should be used, and the first throw should not be pulled so tight that the tissue becomes strangulated. The ends should be cut as short as possible without jeopardizing knot security. Crushing clamps should not be applied to any part of the suture that remains in the tissue as it will weaken the material.

Suture patterns can essentially be divided into *interrupted* and *continuous* patterns; in all interrupted patterns, a knot is tied after each suture is completed, whereas in a continuous pattern, the thread is repeatedly passed through the layers of tissue and only knotted at beginning and end. Obviously, the former is slower but

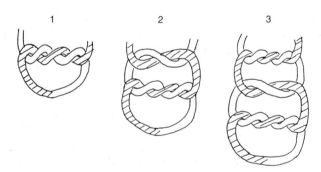

15.2 Knot used with PDS (polydioxanone). The knot is started with a double throw (surgeon's knot) (1). This will hold under moderate tension until locked with a single reverse throw (2). The knot is secured with a final double throw (3). All throws must be snugged down firmly.

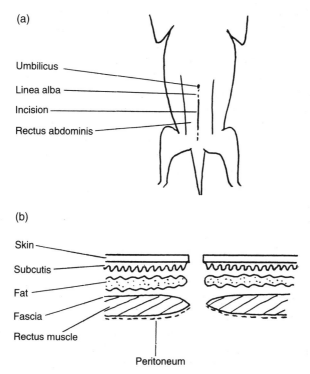

(a)

Umbilicus
Linea alba
Incision
Rectus abdominis

(b)

Skin
Subcutis
Fat
Fascia
Rectus muscle

Peritoneum

15.3 Mid line abdominal incision. (a) Ventral view. (b) Transverse section.

has the advantage of added security—if one suture breaks down, or one knot comes undone, the whole row is not lost. The latter is quicker therefore reducing the time that the animal spends under anaesthesia, but less secure.

To demonstrate the various patterns available, the closure of a midline abdominal incision can be considered (see Fig. 15.3).

In closing such an incision typically 3 (or 4) layers of sutures would be used as follows: muscle, fat, (subcutis), and skin.

1. *Muscle layer*

   When suturing this layer, it is important to remember that the actual holding power of muscle is poor—the layers that surround it are much better (fascia and peritoneum in this case), so when placing sutures in the abdominal body wall, it is vital to ensure the fascia and peritoneum (a silvery layer adherent to the under surface of the muscle) are both included in the suture.

   A *simple interrupted pattern* is preferred here for added security. (See Figures 15.4 a and b.)

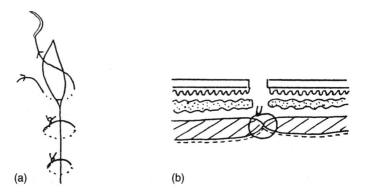

15.4 (a) Simple interrupted suture. (b) Transverse section.

2. *Fat*

   Simple interrupted sutures could also be used here but in practice most people use a *simple continuous* pattern for speed. (See Figure 15.5 a and b).

   It is important that the suture is not pulled tight or fat necrosis will result.

3. *Subcuticular*

   Optional. Essentially a *continuous horizontal mattress* suture in the subcutis. (See Figure 15.6.)

4. *Skin*

   The *simple interrupted* suture is best on most occasions. Alternatively, some people prefer (interrupted) *horizontal mattress* sutures for speed (although this does cause eversion of the wound edges). (See Figure 15.7 a and b.)

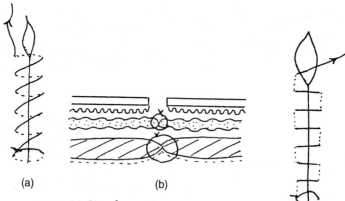

15.5 (a) Simple continuous suture.
(b) Transverse section.

15.6 Subcuticular suture.

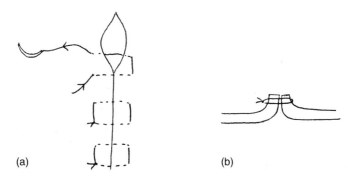

15.7 (a) Horizontal mattress suture. (b) Transverse section.

Vertical mattress sutures can be used if skin is under tension. (See Figure 15.8 a and b.)

SUTURE MATERIALS

There is a wide choice of both type and size of suture materials available. Suture materials can be described as being:

- **natural** or **synthetic**
- **absorbable** or **non-absorbable**
- **monofilament** or **braided**

No one type is suitable for all situations. It is important to consider:

1. the nature and location of the tissues to be sutured. Different tissues have different healing rates—the peritoneum heals within a few days, skin takes 7–10 days, tendons can take a month or more, and fascia has only 70 per cent of its original strength 9 months post-operatively;

2. whether or not the sutures are to be removed. Absorbable suture materials are digested by body enzymes and will eventually dissolve. They are generally used in buried or visceral locations, although some can be used in skin if it is

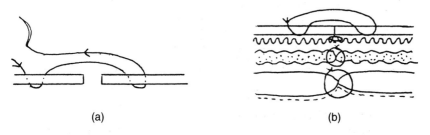

15.8 (a) Vertical mattress suture. (b) Transverse section.

not possible to take them out at a later date (e.g. in a fractious or wild animal).

Non-absorbable suture materials are not digested but remain 'walled off'. They can be used in buried locations such as the body wall, or in skin from where they must be removed post-operatively (typically at about ten days after surgery).

Until the mid 1970s there was little choice of absorbable suture for the surgeon. It was difficult to sterilize sutures, and braided sutures gave rise to post-operative infections more easily than monofilaments. This is because fluids are drawn from the end of the suture into the knot and the tissue by capillary action. However, modern sutures are sterilized by the manufacturer and a better awareness of sterility reduces the risk of infection. Modern synthetic materials are far superior to the old natural products like catgut and silk, and they are becoming the standard ones in use. Synthetic suture materials are stronger, more predictable, and cause much less tissue reaction than natural materials.

Natural absorbable sutures
Catgut is a natural suture material made from the submucosa of sheep intestine or the serosa of bovine intestine. It is available as plain or chromic catgut. The former is very rapidly and unpredictably absorbed, whereas the latter, which has been soaked in chromic salt solution, resists enzyme action and will support tissue for about 10 days. It can be used in all tissue layers except those that require extended tissue holding. Since it is absorbed by cells and enzymes, it has been known to disappear in only 72 hours if implanted in patients with a high white cell count.

Synthetic absorbable sutures
Coated Vicryl* (*Polyglactin 910*) and *Dexon* are both synthetic absorbable braided sutures which are absorbed by simple water hydrolysis and give a predictable 21 days support in tissue. Dexon is a polyglycolic acid. Its absorption is complete in 90–130 days. Vicryl is a copolymer and differs from the polyglycolic acid sutures in that it has two components, glycolide and lactide. Once its holding strength has gone, it is absorbed much more quickly from the tissue and this process is complete in 70 days. Vicryl is coated with polyglactin 370 and calcium stearate, which allow knots to be snugged down easily, and therefore provides more secure knotting. Dexon does not perform so well in this respect.

PDS (polydioxanone) is a monofilament synthetic absorbable suture with high tensile strength which provides wound support for more than 42 days. It has good handling and knotting properties and has been described as being as near to the ideal suture as possible.

Natural non-absorbable sutures
The natural non-absorbable materials are *linen* and *silk*. Both of these are multifilament: the linen is twisted and and the silk is braided. Both give rise to an inflammatory reaction and have been largely superceded by more modern materials.

Synthetic non-absorbable sutures
The oldest synthetic non-absorbable material is *nylon* (polyamide). This is available as monofilament or braided. Larger sizes of mono-filament nylon can be difficult to tie. The braided form is also available with a sheathed coat which improves handling and reduces capillarity. Nylon is commonly used as a skin closure material. It can be used as a buried suture but can cause problems such as sinus formation. *Prolene* is monofilament polypropylene and is also used in skin and as a buried suture. It is totally inert and when buried in tissue is not affected by enzymes or tissue fluid. It is used for vascular surgery and other situations where long-term reliability is required.

Stainless steel wire
This is available as a suture material and is used in special situations such as orthopaedic surgery.

Staples
These may be used for wound closure for the skin layer. They are supplied sterile inside a special applicator and are removed after 10 days by a purpose-made device. Removal can cause mild pain and may be resented by some animals. It is important to be accurate with the apposition of skin edges when using this technique. The main advantage of staples is their speed of application. *Michel clips* are applied using special applicator forceps and have to be sterilized before use. There is a tendency for them to be over-crimped, leading to tissue damage and necrosis which acts as a focus for wound infection. This over crimping cannot occur with skin staples because of the design of the applicator and the staples which allow for the normal post-operative tissue oedema that occurs in the first phase of wound healing.

Tissue adhesives
n-Butyl cyanoacrylate adhesive (Vetbond, 3M) can be used to bond tissue together. On contact with tissue and body fluids it changes from a liquid to a solid state by rapid polymerization and thus seals the wound. It can only be used on relatively small skin

Straight needles

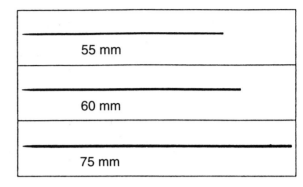

Curved needles

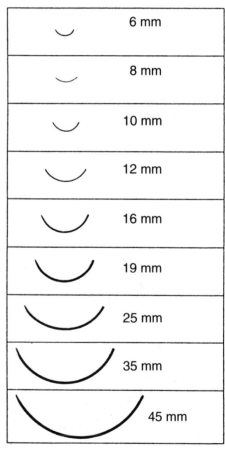

15.9 Needle sizes and shapes.

Half-circle needles

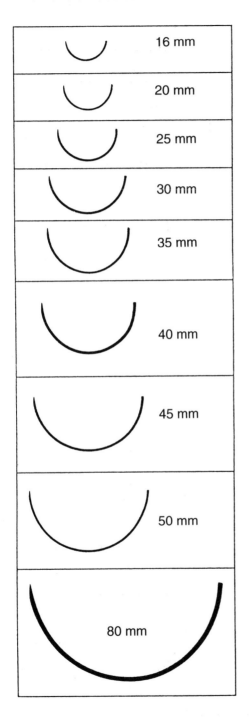

16 mm

20 mm

25 mm

30 mm

35 mm

40 mm

45 mm

50 mm

80 mm

Table 15.1 Comparison of metric and BP/BPC gauges

		Thinnest					
	Metric number	0.1	0.2	0.3	0.3	0.4	0.5
	Catgut						
BP/BPC	Non-absorbables and synthetic absorbables			10/0	9/0	8/0	8/0	7/0

wounds where there is no tension and when there is unlikely to be interference from the animal by excess grooming. It is most useful, not as a replacement for sutures but as a supplement to them, and will help to approximate wound edges between sutures. If used with staples, it should only be placed *between* the staples and not on them, or it may interfere with staple removal.

SUTURE NEEDLES

Surgical needles either have eyes or are swaged on to the suture material in manufacture and are therefore eyeless. When this type of eyeless suture is used only one thickness of the material has to be dragged through the tissue and thus there is much less tissue damage which improves wound healing. It also eliminates the chore of cleaning and resterilizing needles and the surgeon has a new sharp needle each time thus reducing tissue bruising from use of blunt needles.

The needles are either straight or bent into shape as part of a circle (see Figure 15.9) and are measured in millimetres. The degree of curvature is to help the surgeon work where space is limited. Thus in areas where access is unrestricted the straightest needle possible should be used for most accurate suturing.

Needles are either round bodied for suturing soft tissue, such as bowel and other internal organs, or cutting for use in skin or tough fibrous tissue, such as fascia. Curved cutting needles can be conventional, with the apex of the triangular profile of the needle on the inner curvature, or reverse cutting where the apex is on the outer curvature thus reducing the possibility of the suture cutting through the tissue. There is also available a tapercut needle which is a round bodied needle with a tiny cutting tip on the end.

SIZES OF SUTURE

Sutures come in a variety of sizes, denoting the diameter of the material, and in different precut lengths. The metric system, where the diameter is represented in tenths of a millimetre, is gradually

Table 15.1 (cont.)

									Thickest	→
0.7	1	1.5	2	3	3.5	4	5	6	7	8
	6/0	5/0	4/0	3/0	2/0	0	1	2	3	4
6/0	5/0	4/0	3/0	2/0	0	1	2	3 & 4	5	6

replacing the old BP/BPC gauges (see Table 15.1).

The choice of appropriate sized material is governed by the nature of the tissues and the size of the animal. It is important to ensure adequate tensile strength whilst remembering that using too thick a suture material will impair healing.

A rough guide for size (metric numbers—M) is as follows:

Ligatures	Catgut or synthetic absorbable, e.g. Vicryl. (2 M–4 M)
Skin	Nylon (monofilament/braided), Prolene or synthetic absorbable (1.5 M–4 M)
Subcuticular	Catgut or synthetic absorbable (1.5 M–3.5 M)
Muscle	Catgut or synthetic absorbable (2 M–4 M)

PACKAGING

Modern sutures are supplied sterile in a double layered pack, usually in boxes of twelve. The outer envelope has a clear window through which the inner foil pack can be seen and gives the following information:

> Type of suture material
> Gauge of material
> Needle size and type
> Length of material
> Method of sterilization
> Date of manufacture
> Batch number
> Code number for re-ordering

The expiry date is 5 years from manufacture. After this period, absorbable sutures will lose their strength and the time for absorption becomes unpredictable. The cost of the suture is actually very small when compared to the total cost of the experiment, and it is most unwise to use out of date sutures and risk wound breakdown.

16 Useful equipment

....

BLOOD COLLECTION AND INTRAVENOUS ADMINISTRATION OF SUBSTANCES

Since a considerable amount of skill is required to perform intravenous injections competently, it is desirable to gain practical experience on a cadaver or a model prior to performing the procedure on a live animal. The KOKEN rat (B&K Universal Ltd) is an anatomically correct model rat with life-like tail veins on which intravenous injections and blood sampling can be practised. It is also possible to practise gavage and endotracheal intubation, making the KOKEN rat a particularly useful training tool.

Over-the-needle cannulae

Flexible over-the-needle cannulae can be obtained in a variety of sizes to suit the species and the intended site of administration or withdrawal (see Table 16.1). They are easily inserted into the superficial veins of most species (see Chapter 10).

Table 16.1
Recommended sizes for cannulae

Species	Site	Gauge	Length (inches)
Rat	Tail vein	24–25	$1/2$–$3/4$
Rabbit	Ear vein	24	$3/4$
Dog	Cephalic or jugular vein	20–21 *	1–$1^1/2$
Cat	Cephalic or jugular vein	22–23	1
Rhesus monkey	Cephalic or saphenous vein	23–24	$3/4$–1
Pig	Ear vein	21–23	1–$1^1/2$
Sheep/goat	Jugular vein	19–21	$1^1/2$
Cattle/horses	Jugular vein	19–21	$1^1/2$

* Depends on breed and age. For puppies, a 23 or 25 gauge cannula can be used.

Evacuated blood collection tubes

These are tubes with a rubber seal on one end which contain a partial vacuum, for example Vacutainers (Becton Dickinson), and Wexvac tubes or Monovettes (Sarstedt) (See Figure 16.1). Double-ended needles and a special holder are required to use evacuated tubes. The needle is screwed into the holder, with one end of the needle just touching the rubber seal of the tube without penetrating it. The other end of the needle is placed into the vein, and the tube advanced until the needle penetrates the seal. The vacuum in the tube sucks blood into the tube. If the tube is advanced too soon, then air will be sucked into the tube and the vacuum will be lost. Once the tube is full, no more blood is sucked in. The tube should be taken out of the holder as the needle is withdrawn from the animal. Evacuated tubes are quick and easy to use, but are only useful if the vein is large enough to allow rapid flow of blood. Occasionally, the rapid passage into the evacuated tube will damage the red blood cells and result in haemolysis.

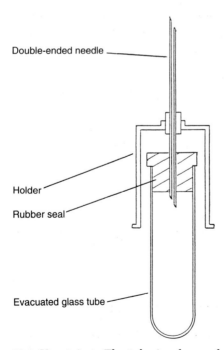

Double-ended needle

Holder

Rubber seal

Evacuated glass tube

16.1 Vacutainer. The tube is advanced so that the rubber seal is penetrated only once the needle is in the vein, otherwise air enters the tube and the vacuum is lost.

Butterfly needles
These are needles with a long flexible tube attached, with a fitting on the end for syringes (see Fig. 10.3). The needle can be taped in place in a vein using the wings, and are useful because a syringe can be attached at some distance from the animal. They can be used for administration of substances and withdrawal of blood.

Tourniquets
When performing intravenous injections or withdrawing blood from some superficial veins, such as the cephalic, jugular or saphenous veins, the vein must be raised in order for a needle to enter it. This can be done by an assistant, placing their thumb over the vein proximal to the site of venepuncture, but this can be awkward, and it is often easier to use a tourniquet. These occlude the venous drainage, raising the vein, and usually have quick release mechanisms so they can be removed rapidly (see Figure 10.1 a and b Vetourniquet, Vet-2-Vet).

Vasodilating agents
For some smaller species, e.g. rabbits and rats, collecting blood can be facilitated by applying vasodilators to a superficial vein to engorge it and increase the blood flow. Warm water has some effect, but there are safe chemical agents available which cause vasodilation within 5 to 10 minutes after application, and which can be wiped off once the blood has been collected. An example is d-limonene oil (Vasolate, IMS). Note that xylene is carcinogenic and its use is not recommended.

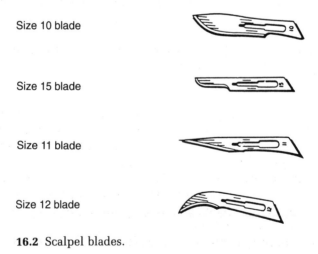

Size 10 blade

Size 15 blade

Size 11 blade

Size 12 blade

16.2 Scalpel blades.

SURGICAL EQUIPMENT

A wide variety of surgical instruments and other equipment is available, and the licensee should consult the many catalogues published by instrument manufacturers to determine which equipment is suitable for their particular project (see Directory).

Scalpel blades (see Chapter 14)

Scalpel blades are available in a variety of sizes, which are designed for different purposes (see Fig. 16.2).

Size 10 blades are all-purpose blades with a rounded cutting edge which can be used for most surgical procedures.

Size 15 blades are similar to size 10 but smaller, and are useful all-purpose blades for fine work on small animals.

Size 11 blades are pointed, with a straight blade, and are particularly useful for work involving fine cutting of connective tissue.

Size 12 blades are curved with the cutting edge on the inside of the curve. They are useful for situations where the skin needs to be cut without damaging the structures beneath.

Suture materials and suture needles (see Chapter 15)

Suture materials are available in a variety of types and sizes, with or without needles attached and there are materials to suit all requirements. Polyglactin 910 (Vicryl, Ethicon) is probably the most versatile absorbable suture. The size required depends on the species and situation. Size 2 or 3 metric Vicryl will be appropriate for small rodents and rabbits respectively and size 3 or 4 for larger species (see Chapter 15 for appropriate needle sizes).

Nylon and Polypropylene (Prolene, Ethicon) are useful non–absorbable sutures. The same or slightly larger sizes can be used as for absorbable sutures.

A tissue adhesive is available, n-butyl cyanoacrylate (Vetbond,3M), which can be used for sealing small wounds.

Useful instruments

Crocodile forceps are long, fine instruments with small jaws, operated like scissors (see Figure 16.3). They are particularly useful for retrieving material from deep within a cavity, and are often used for removing foreign bodies from the ear canal.

Metzenbaum scissors are long, fine scissors with rounded tips (see Figure. 16.4). They are very useful when operating because the handles are long in comparison with the blades, which allow the tip to be precisely controlled.

Blunt, curved on flat scissors are useful general purpose scissors

16.3 Crocodile forceps.   **16.4** Metzenbaum scissors.

(see Fig. 16.5). They are good for clipping fur from small areas, as the curved blades prevent the skin from being cut. They are also useful for cutting dressings, etc., as again the blunt tip prevents trauma to the skin.

Forceps are available in many different patterns (see Figure 16.6). Blunt forceps are useful for dissecting and holding delicate tissues. Forceps with a rat-toothed pattern are more traumatic to use, and are used for tough tissues such as skin.

Retractors are used to hold tissues apart to improve surgical access. Gelpi retractors have small hooks on the end which are inserted between the tissues to be separated (see Figure 16.7). As the handles are pushed together, the hooks separate forcing the tissues apart. There is a ratchet to hold the retractor open.

Needle holders come in many patterns (see Figure 16.8). Gillie's needle holders combine a cutting blade with jaws to hold a needle, so

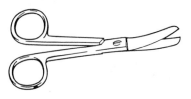

16.5 General purpose scissors.

(a)

(b)

16.6 Forceps. (a) Blunt Forceps; (b) Rat-toothed Forceps.

16.7 Gelpi retractors.

separate scissors are unnecessary. They do not have a ratchet, however, so considerable digital pressure may be required to hold the needle if the tissue is tough. Right and left-handed versions are available.

Olsen–Hegar needle holders are like Gillie's but have a ratchet, and are particularly useful.

Mayo needle holders are similar to Olsen–Hegar's but have no cutting blade.

MacPhail's needle holders have an unique handle which does not require the fingers and thumb to be placed through metal rings. Much practice is required to be competent with them, but they are comfortable to use for long periods of time.

Artery forceps are used to clamp bleeding vessels for haemostasis. Spencer–Wells and Dunhill artery forceps are general purpose haemostats (see Figure 16.9). Mosquito forceps are very fine, and are useful for small animals and for delicate work. Kocher's artery forceps combine a rat-toothed tip with crushing jaws, and are useful for grasping bleeding vessels deep within tissues.

Allis tissue forceps are long forceps with small gripping jaws (see

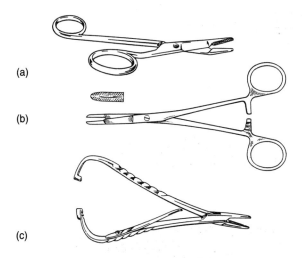

(a)

(b)

(c)

16.8 Needle holders. (a) Gillie's needle holders; (b) Olsen–Hegar's needle holders; (c) MacPhail's needle holders.

(a)

(b)

(c)

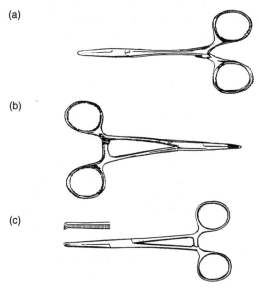

16.9 Artery Forceps. (a) Spencer–Wells Forceps; (b) Dunhill's Forceps; (c) Kocher's Forceps.

Figure 16.10). They are useful for grasping pieces of tissue to hold them to one side temporarily, and cause relatively little trauma.

POST-OPERATIVE CARE EQUIPMENT
Bedding
Artificial sheepskin bedding is warm and comfortable, keeps animals dry, and avoids the problem of sawdust sticking to moist areas such as the mouth, nose, perineum or wound. Examples are Vetbed, (Cox), and polyester fibre bedding (Veterinary Drug).

Food and water
Animals recovering from anaesthetic may be unable or unwilling to move across the cage to reach their food and water. Food can

16.10 Allis tissue forceps

be placed alongside them relatively easily, but water is harder to supply. Shipping diets (e.g. Transgel, Charles River) have a high water content but are solid, so they can be placed in the cage with the animal to avoid the need to walk around. Reconstituted fruit-flavoured jelly can serve the same purpose.

For some species, it is advisable to use convalescent diets in the immediate post-operative period. Pedigree Petfoods and Hills Pet Products both make convalescent diets for dogs and cats.

Heating
To prevent post-operative hypothermia, extra heat may need to be supplied, with care being taken not to burn the animal's skin. Electronic heat pads are available (e.g. from IMS) with thermostats linked to the animal's body temperature. Reusable instant heat pouches, for example, Safe and Warm (Trident Pharmaceuticals, or Arnolds Veterinary Products) provide heat anywhere without the need for electricity. The pouch must be placed within an insulating bag to prevent excess heat reaching the animal. These can also be used for warming fluids prior to injection.

Wound dressings and haemostasis
Good surgical technique should prevent excess bleeding from a surgical site, but if haemorrhage occurs, there are several ways to arrest it. Digital pressure for a few minutes is sufficient for most cases. Pressure bandages may be applied to the limbs or even to the trunk in larger animals, with care being taken not to make the bandage too tight. Bandages should have three or four layers. The layer adjacent to the wound should be a non-stick dressing, such as Melolin (Smith and Nephew) or Rondopad (Millpledge), which allow fluids to pass through into the next layer, or an occlusive dressing which prevents the passage of fluid into the next layer, keeping the wound moist. The latter is useful for large contaminated wounds and for promoting epithelialization. The second layer should be a thick layer of absorbent material, such as Soffban (Smith and Nephew) or Vet-Orthoband (Millpledge), which are conforming and comfortable. Cotton Wool is not ideal as it tends to bunch up and create areas of increased pressure under the outer wrapping. The third layer should be a fairly tight wrapping of conforming bandage such as Vet-K-band (Millpledge) or Kling bandage. Finally, a layer of elastic bandage such as Elastoplast (Smith and Nephew), Vetrap (3M), or Co-ripwrap (Millpledge) is used to hold the dressing in place.

In some cases, for example if there is prolonged haemorrhage, haemostatic dressings may be required, e.g. Xenocol (Ichor). These are applied to the surface of the wound, and assist in arresting haemorrhage.

MISCELLANEOUS EQUIPMENT
Clippers
As animals have hair this must be shaved before surgery can be performed. Clippers are available with a variety of blades for different purposes. All-purpose blades can be used on most species, and there are special blades for coarse hair and for clipping sheep. Size 40 blades are general blades for surgical and laboratory use. Size 10 blades are useful for sensitive skin and in awkward areas.

Clippers need to be maintained in good condition, or they stop clipping effectively. They should be brushed after each use, to remove the loose hair, and sprayed with a disinfecting lubricant spray such as Clippercide. If this is not done, they start to pull the hair and can cut the skin, particularly in animals with delicate skin such as rabbits. Periodically, the blades need to be sharpened.

Bowls and dishes
Kidney dishes are particularly useful for holding solutions or materials. They can be used as wastebins for the disposal of used swabs or tissue debris, etc., and a selection of different sized kidney dishes is very useful in the operating theatre. They are available in materials which can be autoclaved.

Anaesthetic machines (see Chapter 12)
It is essential to have an anaesthetic machine with an appropriate vaporizer for long-term inhalation anaesthesia. These allow precise control of the amount of volatile anaesthetic agent delivered to the animal, and therefore allow precise control of the depth of anaesthesia. They are available with varying degrees of sophistication, from an oxygen cylinder on a trolley with a vaporizer to a workstation with multiple gas cylinders and facilities for forced ventilation.

Endotracheal tubes (see Chapter 12)
These are available in many different sizes for different species, with or without inflatable cuffs. The lubricated tube should be passed over the tongue and epiglottis, and through the larynx. Once the tube is *in situ*, the cuff is inflated to seal the trachea and prevent the animal from being able to breathe around the tube. For tubes smaller than 5 mm diameter, inflatable cuffs should not be used, as these reduce the size of the airway, and in small animals this is undesirable.

The size of tube used depends on the size of the animal, but as a rough guide, adult beagles will need a 9 or 10 mm tube, with a cuff, cats need a 3–5 mm tube without a cuff, and rats need a modified urinary catheter. Pigs have a particularly long larynx, and a laryngoscope is required for endotracheal intubation. This illuminates the larynx and also holds down the tongue and epiglottis. Rabbits can be intubated via surgical implantation of an endotracheal tube in the trachea in non-recovery situations, or by passing a tube carefully into the mouth. The tube is inserted in the mouth to one side of the incisors, as far as the glottis. The tube is then advanced into the larynx on the inspiratory phase of respiration (Conlon *et al* 1990, Alexander and Clark 1980).

Directory of useful names and addresses

Manufacturers and suppliers

Alfred Cox (Surgical) Ltd
Edward Road
Coulsdon
Surrey CR3 2XA

Arnolds Veterinary Products Ltd.
Cartmel Drive
Harlescott
Shrewsbury
Shropshire SY1 3TB

B & K Universal Ltd
The Field Station
Grimston
Aldborough
Hull
North Humberside HU 11 4QE

Beta Medical and Scientific
(Datesand Ltd)
2 Ferndale Road
Sale
Cheshire M33 3GP

Blease Medical Equipment Ltd
Deansway
Chesham
Bucks. HP5 2NX

Boehringer Ingelheim Ltd
Animal Health Division
Ellesfield Avenue
Bracknell
Berkshire G12 4YS

Bowring Medical Engineering Ltd
Unit F, New Yatt Business Centre
New Yatt
Witney
Oxon. OX8 6TJ

Brookwick, Ward and Co. Ltd
8 Shepherds Bush Road
London W6 7PQ

Catac Products
Catac House
1 Newnham Street
Bedford MK40 3TR

Centaur Services
Centaur House
Torbay Road
Castle Cary
Somerset BA7 7EU

Charles River UK Ltd
Manston Road
Margate
Kent CT9 4LT

Ethicon Ltd
PO Box 408,
Bankhead Avenue
Edinburgh EH11 4HE

Harvard Apparatus Ltd
Fircroft Way
Edenbridge
Kent

Hills Pet Products Ltd
Sherwood House
33/35 Wellfield Road
Hatfield
Herts. AL10 OBS

Ichor Ltd
41 Wheatsheaf Road
Alconbury Weston
Huntingdon
Cambs. PE17 5LF

International Market Supply
Dane Mill
Broadhurst Lane
Congleton
Cheshire CW12 1LA

Isotec Ltd
Shaws Farm
Blackthorn
Bicester
Oxon. OX6 OTP

Kontron Instruments
Blackmoor Lane
Croxley Centre
Watford
Herts. WD1 8XQ

Lab Care Precision Ltd
Calleywell Barn
Goldwell Lane
Aldington
Kent TN25 7DX

W M Lillico and Son Ltd
Wonham Mill
Betchworth
Surrey RH3 7AD

MDC
Hamelin House
211/212 Hightown Road
Luton
Beds. LU2 OVZ

Millpledge Pharmaceuticals Ltd
Whinleys Estate
Church Lane
Clarborough
Retford
Notts. DN22 9NA

Modular Systems and
Development Company Ltd
Campwood Road
Rotherwas Industrial Estate
Hereford HR2 6JD

Moredun Animal Health Ltd
408 Gilmerton Road
Edinburgh EH17 7JH

NKP Cages Ltd
1 Bilton Road
Erith
Kent DA8 2AN

Pedigree Petfoods and Waltham
Centre for Pet Nutrition
Waltham-on-the-Wolds

Melton Mowbray
Leics. LE14 4RT

Penlon Ltd
Radley Road
Abingdon
Oxford OX14 3PH

Portex Ltd
Hythe
Kent CT21 6JL

Rocket of London Ltd
Imperial Way
Watford WD2 4XX

RS Biotech
Tower Works
Well Street
Finedon
Northants. NN9 5JP

Sandown Scientific
11 Copsem Drive
Esher
Surrey KT10 9HO

Shor-Line Ltd
Unit 39A/39B
Llandow Industrial Estate
Cowbridge
South Glamorgan CF7 7VY

Silogic Design Ltd
Enterprise House
181–189 Garth Road
Morden
Surrey SM4 4LL

Smith and Nephew Medical
Ltd
PO Box 81, Hessle Road
Hull HU3 2BN

Southern Veterinary Services
Brooks Road
Lewes
East Sussex BN7 2AL

Special Diets Services
PO Box 705
Witham
Essex CM8 3AD

Spillers
PO Box 53
Newmarket
Suffolk CB8 8QF

Trident Pharmaceuticals Ltd
Brooks Road
Lewes
East Sussex BN7 2AL

UAR
PO Box 9
Huntingdon
Cambs. PE17 5LA

Veterinary Drug Co PLC
Derwent Valley Industrial Estate
Dunnington
York YO1 5RS

Veterinary Instrumentation
50 Broomgrove Road
Sheffield S10 2NA

Vet-2-Vet Marketing
PO Box 98
Bury St. Edmunds
Suffolk IP33 2QN

Willington Medicals Ltd
Govan Road
Fenton Industrial Estate
Stoke-on-Trent ST4 2SZ

Wyvern Veterinary Company Ltd
Bulwark Industrial Estate
Chepstow
Gwent NP6 5QZ

Wholesalers

Alfred Cox (Surgical) Ltd.

Arnolds Veterinary Products Ltd

Centaur Services

Millpledge Pharmaceuticals Ltd

Southern Veterinary Services

Veterinary Drug Co PLC

Willington Medicals Ltd

These companies, and others,
market a wide range of products
for veterinary use.

Surgical equipment

All the wholesalers listed above
market a range of veterinary instruments and surgical equipment.
In addition, Rocket of London
and Veterinary Instrumentation
also supply surgical instruments.
Ethicon Ltd are manufacturers and
suppliers of suture materials and
skin-stapling devices. Vet-2-Vet
Marketing supply a range of
gadgets including the quick release
tourniquet.

Anaesthetic equipment and related apparatus

Alfred Cox (Surgical) Ltd

Arnolds Veterinary Products Ltd

Blease Medical Equipment Ltd

Brookwick, Ward and Co Ltd

International Market Supply

Kontron Instruments

Penlon Ltd

Portex Ltd

Veterinary Drug Co PLC

These companies manufacture or
market anaesthetic machines, va-
porizers, circuits, induction chambers, face masks, endotracheal
tubes, and monitoring equipment.

Monitoring equipment

Alfred Cox (Surgical) Ltd
Respiration monitors

Bowring Medical Engineering Ltd
ApneAlarm respiration monitor

Hillmoore Electronic Consultants
22 High St,
Hanslope,
Milton Keynes MK19 7LQ
'Zoovent' animal ventilator

Silogic Design Ltd
Electrocardiogram

Veterinary Instrumentation
Respiration monitors

Restraining devices
Alfred Cox (Surgical) Ltd
Dog and cat catchers, nets, crush
cages, sheep restrainers, muzzles,
and Elizabethan collars. Suppliers
of Buster X-ray Support troughs
and Buster Vacu-Support bags for
easy positioning of animals under
anaesthetic

Harvard Apparatus Ltd
Perspex boxes, harnesses,
and tethers

International Market Supply
Rodent and rabbit holders, primate
chair, nets, bags, muzzles and gags.
Two sizes of operation support
for holding animals in the correct
position for surgery

MDC
Muzzles, traps, cat catchers and
nets, and long-reach hypodermics

**Housing and husbandry-related
equipment**
Housing
B & K Universal Ltd
Rodent and rabbit cages

Isotec Ltd
Isolators

Lab Care Precision Ltd
Rabbit, small rodent, guinea-pig,
ferret, and cat cages

W M Lillico and Son Ltd
'Scantainer' ventilated, airfiltered
multipurpose cabinet for isolating
animals

Modular Systems and
Development Co Ltd
Versatile primate cages

Moredun Animal Health Ltd
Isolators

NKP Cages Ltd
Rat, mouse, guinea-pig, and
rabbit cages

Shor-Line Ltd
Dog and cat cages

Food
Hills Pet Products Ltd
Dog and cat foods, and
convalescent diets

Pedigree Petfoods and Waltham
Centre for Pet Nutrition
Dog and cat foods, and
convalescent diets

Special Diets Services
Diets for all animals

Spillers
Dog and cat foods

UAR
Diets for laboratory animals

Bedding
Alfred Cox (Surgical) Ltd
Vetbed artificial sheepskin

Beta Medical and Scientific
(Datesand Ltd)
Aqua-sorb tray liners, bedding,
Virkon virucidal disinfectant

W M Lillico and Son Ltd
Litalabo for rodents

Modular Systems
Tray Liners

RS Biotech
Rodent and rabbit bedding

Special Diets Services
Rodent and rabbit bedding

Wyvern Veterinary Company Ltd
Flectabed thermally insulated
pads

Environmental enrichment equipment

Beta Medical and Scientific (Datesand Ltd)
Primate foraging boards

The Company of Animals
PO Box 23 Chertsey
Surrey KT16 OPU
Kong toys and Boomer balls

Post-operative care (see also Bedding)

Arnolds Veterinary Products
Safe and warm reusable heat pouches

Boehringer Ingelheim Ltd
Haemovet haemostatic dressings

Catac Products
Thermostatic heat pads, infra-red units, bedding

Charles River UK Ltd
Transgel shipping diet

Ichor Ltd
Xenocol Haemostatic dressing

International Market Supply
Heat pads, including thermo-statically controlled pads with rectal probes for body temperature measurement

Millpledge Pharmaceuticals
Dressing and bandages

Smith and Nephew Medical Ltd
Bandages

Trident Pharmaceuticals Ltd
Safe and Warm reusable heat pouches

Equipment for euthanasia

International Market Supply
Carbon dioxide chambers

Sandown Scientific
DecapiCone for euthanasia of small rodents

Protective clothing and respiratory protection equipment

3M Occupational Health and Environmental Safety
3M United Kingdom PLC
3M House
Bracknell
Berks. RG12 1JU
Face masks and respirators for all situations

Beta Medical and Scientific (Datesand Ltd)
Protex filtering masks

Brindus (Industrial Services) Ltd.
Dartford Trade Park
Powder Mill Lane
Dartford
Kent DA1 1NQ
All forms of protective clothing including respirators

Pureflo Safety Ltd
Moat House
Wheathampstead
St. Albans
Herts. AL4 3QT
Pureflo powered respirators

Racal Safety Ltd
Beresford Avenue
Wembley
Middlesex HAO 1QJ
Respirators for all situations

LABORATORIES
Diagnostic tests
Vet Diagnostics Ltd
Victoria House
Small Dole
Henfield
West Sussex BN5 9XE
Covers common and exotic species, particularly primate screening

The Microbiology Laboratories
56 Northumberland Road
North Harrow
Middlesex HA2 7RE
Specialize in exotic and laboratory animal microbiology and medicine

Vetlab Services
Unit 11
Station Road
Southwater
Horsham
West Sussex RH13 7ZA
Common species

Histopathology
Finn UK Ltd
The Grange
Weybread
Diss
Norfolk IP21 5TU

Rest Associates
24 Lower End
Swaffham Prior
Cambridge CB5 OHT

Health monitoring
Charles River UK Ltd
All laboratory species, particularly
primates, and kits for in-house use

Harlan Olac
Shaw's Farm
Blackthorn
Bicester
Oxon. OX6 OTP
All laboratory species

Laboratory Animal Unit
Royal Veterinary College
Hawkshead Campus
North Mymms
Hatfield
Herts. AL9 7TA
All laboratory species as well as
domestic animals

The Microbiology Laboratories
56 Northumberland Road
North Harrow
Middlesex HA2 7RE
Post-mortem examinations,
serology, parasitology,and
microbiology

Organon-Technika
Science Park
Milton Park
Cambridge CB4 4BH
Kits for in-house use

Central Public Health Laboratory
Virus Reference Laboratory
61 Colindale Avenue
London NW9 5HT
Serological screening

Analysers for in-house use
Sanofi Animal Health Ltd
PO Box 209, Rhodes Way
Watford
Herts. WD2 4QE
Vet Test 8008 biochemical analyser

Vetlab Services
The Analyst biochemical analyser
QBC-U haematology analyser

OTHER USEFUL ADDRESSES

British Laboratory Animal
Veterinary Association
c/o British Veterinary Association
7 Mansfield Road
London W1M OAT

Health and Safety Executive
HSE Information Centre
Broad Lane
Sheffield S3 7HQ

Home Office
E Division
50 Queen Anne's Gate
London SWlH 9AT

Institute of Animal Technology
5 South Parade
Summertown
Oxford OX2 7JL

Laboratory Animal Science
Association
20 Queensberry Place
London SE7 2DZ

Ministry of Agriculture, Fisheries
and Food
Hook Rise South
Tolworth
Surbiton
Surrey KT6 7NF

Research Defence Society
58 Great Marlborough Street
London W1V 1DD

Royal College of Veterinary
Surgeons
32 Belgrave Square
London SW1X 8QP

Royal Society for the Prevention of
Cruelty to Animals
Causeway
Horsham
Sussex RH12 HIG

Universities Federation for
Animal Welfare
8 Hamilton Close
South Mimms
Potters Bar
Herts. EN6 3QD

Further reading

You are obliged to read the **Guidance on the Operation of the Animals (Scientific Procedures) Act 1986** (HMSO (1990) HC 182) before applying for a licence (referred to as 'the Guidance' in Home Office notes).

The Home Office Code of Practice for the Housing and Care of Animals used in Scientific Procedures (1989) should also be examined carefully.

Health and Safety Commission Management of Health and Safety at Work Approved Code of Practice (HMSO (1992)). This is also mandatory reading.

Journals

Animal Welfare. Universities Federation for Animal Welfare, Potters Bar, Herts. EN6 3QD.

ILAR News. Institute of Laboratory Animal Resources, Washington DC 20418.

Lab Animal. Nature Publishing Co, New York NY 0012-2467.

Laboratory Animals. Royal Society of Medicine Services, London W1M 8AE.

Animal Technology. The journal of the Institute of Animal Technology.

Laboratory Animal Science. American Association for Laboratory Animal Science. Cordova TN 38018.

Videos and audio-visual programmes

Anatomy of small mammals. Rijksuniversiteit de Utrecht, Fakulteit der Diergeneeskund, Vakgroep Funktionele Morfologie, Postbus 80. 157, 3508 TD Utrecht, The Netherlands.

Animals in science teaching (1988). British Universities Film and Video Council and Universities Federation for Animal Welfare, Potters Bar, Herts. EN6 3QD.

Animal care training
   DISK 1.   Health and safety
   DISK 2.   Animal handling

DISK 3. Principles of anaesthesia
DISK 4. Principles of surgery
DISK 5. Post-operative care
DISK 6. Euthanasia and carotid cannulation
DISK 7. A Caring Act
DISK 8. Dosing and sampling
DISK 9. Signs of pain and distress
DISK 10. Experimental design
Association of British Pharmaceutical Industry London SWIA 2DY, interactive videos.

Monitoring general anaesthesia in dog and cat (1987). The Unit for Veterinary Continuing Education, The Royal Veterinary College, Royal College Street, London NWI OTU.

Textbooks

Fowler, M. (ed.) (1986). *Zoo and wild animal medicine.* W. B. Saunders, London.

Guide to care and use of experimental animals (1980) Vol. 1. Canadian Council on Animal Care, Ottawa, Ontario.

Guide to care and use of experimental animals (1984) Vol. 2. Canadian Council on Animal Care, Ottawa, Ontario.

Guidelines on the care of laboratory animals and their use for scientific purposes
    Vol. I. (1987). Housing and care
    Vol. II (1989). Pain, analgesia and anaesthesia
    Vol. III. (1989). Surgical procedures
    Vol. IV. (1990). Planning and design of experiments.
Laboratory Animal Science Association, London and Universities Federation for Animal Welfare, South Mimms.

Harkness, J. E. and Wagner, J. E. (1983). *The biology and medicine of rabbits and rodents*, (2nd edn). Lea & Febiger, Philadelphia.

Inglis, J. K. (1980) *Introduction to laboratory animal science and technology.* Pergamon Press, Oxford.

Lane, D. R. (ed.)(1989). *Jones's animal nursing*, (5th edn). British Small Animal Veterinary Association Cheltenham.

Poole, T. B. (ed.) (1986). *The UFAW handbook on the care and management of laboratory animals*, (6th edn). Longman, Harlow, Essex

Tuffery, A. A. (ed.) (1987). *Laboratory animals. An introduction for new experimenters.* John Wiley, Chichester.

Chapter 1   Introduction: training and the personal licensee

LASA recommendations on education and training for licence holders under the UK Animals (Scientific Procedures) Act 1986—FELASA categories B and C. Report of a Committee accepted by LASA Council June 1992. *Laboratory Animals* (1993). **27**(3), 189–206.

Smith, M. W. (ed) (1984). Report of the working party on courses for animal licensees. *Laboratory Animals*, **18**, 209–20.

Chapter 2   Legislation and ethical considerations

Bateson, P. (1986). When to experiment on animals. *New Scientist*, 20 February.

Bunyan, J. (1991). *Handbook for the animal licence holder*, (2nd edn). Institute of Biology, London.

Carruthers, P. (1992). *The animals issue—Moral theory in practice*. Cambridge University Press.

CITES, US Department of the Interior. *Convention on international trade in endangered species of wild fauna and flora. Federal Register*, **42** (35), 10 462–88.

Cooper, M. E. (1981). *The law for biologists*. Institute of Biology, London.

Cooper, M. E. (1987). *Introduction to animal law*. Academic Press, London.

Council of Europe, Strasbourg (1986). *European convention for the protection of vertebrate animals used for experimental and other scientific purposes*. HMSO, London.

Dawkins, M. S. (1980). *Animal suffering, the science of animal welfare*. Chapman and Hall, London.

Dawkins, M. S. and Gosling, M. (1992). *Ethics in research on animal behaviour*. Academic Press/Association for the Study of Animal Behaviour and the Animal Behaviour Society, London.

Hampson, J. (1985). *Laboratory animal protection laws in Europe and North America*. RSPCA, Horsham, Sussex.

HMSO. *Statistics of scientific procedures on living animals*. Published annually. HMSO, London.

Hollands, C. (1986). *The Animals (Scientific Procedures) Act 1986. The Lancet*, **ii**, 32–3.

Hughes, T. I. (ed.) (1990). *Bio-ethics 1989*. A report of the proceedings of an international symposium on the control of the use of animals in scientific research. Animal Welfare Foundation of Canada, Ontario, Canada.

Jasper, J. M. and Nelkin, D. (1992). *The animals rights crusade: The growth of moral protest*. The Free Press, New York.

Kuchel, T. R., Ros, M., and Barrell, J. (ed.)(1992). *Animal pain: ethical and scientific perspectives*. Australian Council for the Care of Animals in Research and Teaching. Universities Federation for Animal Welfare, Potters Bar, Herts.

Langley, G. (ed). (1989). *Animal experimentation: the consensus changes*. Macmillan, Basingstoke.

Leahy, M. P. T. (1991). *Against liberation: putting animals in perspective*. Routledge, London.

Littlewood, Sir S. (1965). *Report of the departmental committee on experiments on animals*, Cmnd 2641. HMSO, London.

Live Animal Regulations (July 1992), (19th edn). International Air Transport Association, Montreal, Geneva.

Medical Research Council (1993). *Responsibility in the use of animals in medical research*. MRC Ethics Series. Medical Research Council, Cambridge.

Ministry of Agriculture, Fisheries and Food. (1991). *Operations on farm animals*. Ministry of Agriculture, Fisheries and Food, Surbiton, Surrey.

Ministry of Agriculture, Fisheries and Food (1992). *Summary of the law relating to farm animal welfare*. Ministry of Agriculture, Fisheries and Food, Surbiton, Surrey.

Paterson, D. and Palmer, M. (ed.) (1989). *The status of animals, ethics, education and welfare*. CAB International, Wallingford, Oxon.

Paton, W. (1993). *Man and mouse: animals in medical research*. Oxford University Press.

Philips, M. T. and Sechzer, J. A. (1989). *Animal research and ethical conflict: an analysis of the scientific literature 1966-1986*. Springer-Verlag, New York.

Poole, T. (ed). (1987) Legislation and laboratory animals. In *The UFAW handbook on the care and management of laboratory animals*, (6th edn), pp. 99–106. Longman, Harlow, Essex.

Rachels, J. (1990). *Created from animals. The moral implications of Darwinism*. Oxford University Press.

Regan, T. and Singer P. (ed.) (1976). *Animal rights and human obligations*. Prentice-Hall, Englewood Cliffs.

Rhodes, P. (1985). *An outline history of medicine*. Butterworth, Sevenoaks, Kent.

Rollin, B. E. (1989). *The unheeded cry: animal consciousness, animal pain and science*. Oxford University Press.

Rowan, A. N. (1989). *Of mice, models and men. A critical evaluation of animal research*. State University of New York Press, New York.

Royal College of Veterinary Surgeons. *Legislation affecting the veterinary profession in the United Kingdom*. Royal College of Veterinary Surgeons, London.

Rupke, N. A. (ed)(1987). *Vivisection in historical perspective*. Croom Helm, London.

Russell, W. M. S. and Burch, R. L. (1959). *The principles of humane experimental technique*, (Special Edition, UFAW 1992). Universities Federation for Animal Welfare, Potters Bar, Herts.

Singer, P. (1976). *Animal liberation*, (2nd edn). Jonathan Cape, London.

Smith, J. A. and Boyd, K. M. (1991). *Lives in the balance: the ethics of using animals in biomedical research.* Oxford University Press.

Smyth, D. H. (1977). *Alternatives to animal experiments.* Scolar Press, London.

Statistics of scientific procedures on living animals Published annually. HMSO, London.

Chapter 3   Roles and responsibilities of 'named' persons.

Institute of Animal Technology (1987). *Guidance notes for persons specified in certificates of designation as responsible for the day-to-day care of protected animals kept at designated establishments under the Animals (Scientific Procedures) Act 1986.* Institute of Animal Technology, Oxford.

Royal College of Veterinary Surgeons (1993). *Code of practice for named veterinary surgeons.* Royal College of Veterinary Surgeons, London.

Chapter 4   Health, safety, and security

Advisory Committee on Dangerous Pathogens (1990). *Categorisation of pathogens according to hazard and categories of containment.* HMSO, London.

Advisory Committee on Dangerous Pathogens and Advisory Committee on Genetic Modification (1990). *Vaccination of laboratory workers handling vaccinia and related poxvirus infections for humans.* HMSO, London.

Association of the British Pharmaceutical Industry (1987). *Advice on laboratory animal allergy.* Association of the British Pharmaceutical Industry, London.

Bland, S. M., Evans, R., and Rivera, J. C. (1987). Allergy to laboratory animals in health care personnel. *Health problems of health care workers,* (ed. E. A. Emmett), *State of the art reviews,* Vol. 2(3). Hanley & Belfus, Philadelphia.

Collins, C. H. (1985) ed. *Safety in biological laboratories,* Institute of Biology. John Wiley, Chichester.

Collins, C. H. (ed.) (1988). *Safety in clinical and biomedical laboratories.* Chapman and Hall, London.

Green, C. J. (1981). Anaesthetic gases and health risks to laboratory personnel: a review. *Laboratory Animals,* **15**, 397–403.

Health and Safety Commission (1993). *Control of substances hazardous to health. General approved code of practice,* (4th edn). HMSO, London.

Health and Safety Commission (1992). *Health and safety in animal facilities.* Education Services Advisory Committee. HMSO, London.

Health and Safety Commission (1992). *Management of health and safety at work approved code of practice.* HMSO, London.

Health and Safety Commission (1990). *What you should know about allergy to laboratory animals.* Education Services Advisory Committee. HMSO, London.

Health and Safety Executive (1992). *Successful health and safety management,* (HS (G) 65). HMSO, London.

Health and Safety Executive (1990). *The essentials of health and safety at work.* HMSO, London.

Hunskaar, S. and Fosse, R. T. (1993). Allergy to laboratory mice and rats: a review of its prevention, management and treatment. *Laboratory Animals,* **27**(3), 206–22.

Industrial Injuries Advisory Council (1986). *Occupational asthma,* Cmnd 9717. HMSO, London.

Kibby, T., Power, G., and Croner, J. (1989). Allergy to laboratory animals: a prospective sectional study. *Journal of Occupational Medicine,* **31**, 842–6.

Medical Research Council (1990). *The management of simians in relation to infectious hazards to staff.* Medical Research Council Simian Virus Committee, London.

Perkins, F. T. and O'Donoghue, P. N. (ed.) (1969). Hazards of handling simians. In *Laboratory animal handbooks, No. 4.* Laboratory Animals Ltd, London

Royal Society of Chemistry (1992). *Hazards in the chemical laboratory,* (ed. S. G. Luxon), (5th edn). Royal Society of Chemistry, London.

Seamer, J. H. and Wood, M. (ed.)(1981). *Safety in the animal house,* (2nd edn). In *Laboratory animal handbooks, No. 5.* Laboratory Animals Ltd, London.

Smith, M. W. (1987). Safety. In *UFAW handbook on the care and management of laboratory animals,* (6th edn), pp. 170–86. Longman, Harlow, Essex.

Whitley, R. J. (1990). Cercopithecine herpes virus I(B virus). In *Virology,* (ed. B. N. Fields *et al.,* (2nd edn).) Raven Press, New York.

Zangwill, K. M., Hamilton, D. H., Perkins, B. A., Regnery, R. L., Plikaytis, B. D., Hadler, J. L. *et al.* (1993). Cat scratch disease in Connecticut. Epidemiology, risk factors and evaluation of a new diagnostic test. *New England Journal of Medicine,* **329**(1), 8–13.

Chapter 5   Humane methods of killing

Commission of the European Communities (May 1993). *Recommendations for euthanasia of laboratory animals.* Final Report of the Commission of the European Communities, Brussels.

Shores, H. H. (1976). *Animal communication by pheromones.* Academic Press, New York.

Chapter 6 Introduction to laboratory animal husbandry

Bhatt, P. N., Jacoby, R. O., Morse, H. C., and New, A. E. (ed.)(1986). *Viral and mycoplasmal infections of laboratory rodents: effects on biomedical research.* Academic Press, Orlando.

Brown, L. (ed.) (1993). *Aquaculture for veterinarians: fish husbandry and medicine.* Pergamon Press, Oxford.

Clarke, H. E. *et al.* (1977). Dietary standards for laboratory animals: report of the LAC Diets Advisory Committee. *Laboratory Animals,* **11**, 1–28.

Clough, G. (1982). Environmental effects on animals used in biomedical research. *Biological Review,* **57**, 487–523.

Clough, G. (1984). Environmental factors in relation to the comfort and well-being of laboratory rats and mice. In *Standards in laboratory animal management,* pp. 7–24. Proceedings of a LASA/UFAW Symposium. Universities Federation for Animal Welfare, Potters Bar, Herts.

Clough, G. (1987). The animal house; design, equipment and environmental control. In *The UFAW handbook on the care and management of laboratory animals,* (6th edn), pp. 108–43. Longman, Harlow, Essex.

Coates, M. E. (1984). Sterilization of diet. *The germfree animal in biomedical research,* (ed. M. E. Coates and B. E. Gustafssen). In *Laboratory animal handbooks, No. 9.* Laboratory Animals Ltd, London.

Fletcher, J. L. (1976). Influence of noise on animals. Control of the animal house environment, (ed. T. McSheehy). In *Laboratory animal handbooks, No. 7,* pp. 61–62. Laboratory Animals Ltd, London.

Foster, H. L., Small, J. D., and Fox, J. G. (ed.)(1983). *The mouse in biomedical research,* Vol. III. Academic Press, New York.

Gamble, M. R. (1982). Sound and its significance for laboratory animals. *Biological Review* **57**, 395–421.

Greenman, D. L., Bryant, P., Kodell, R. L., and Sheldon, W. (1982). Influence of cage shelf level on retinal atrophy in mice. *Laboratory Animal Science,* **32**, 440–50.

Hawkins, A. D. (ed.)(1981). *Aquarium systems.* Academic Press, London.

King, J. O. L. (1978). *An introduction to animal husbandry.* Blackwell, Oxford.

Laboratory Animal Breeders Association of Great Britian Ltd (LABA) and Laboratory Animal Science Association (LASA) (1993). Guidelines for the care of laboratory animals in transit. *Laboratory Animals,* **27**(3), 93–107.

Ministry of Agriculture, Fisheries and Food (1993). *Codes of recommendations for the welfare of livestock: Cattle.* Ministry of Agriculture, Fisheries and Food, Department of Agriculture and Fisheries for Scotland, Welsh Office Agriculture Department, MAFF Publications, London.

Ministry of Agriculture, Fisheries and Food (1990). *Codes of recommendations for the welfare of livestock: Goats.* Ministry of Agriculture, Fisheries and Food, Department of Agriculture and Fisheries for Scotland, Welsh Office Agriculture Department, MAFF Publications, London.

Ministry of Agriculture, Fisheries and Food (1990). *Codes of recommendations for the welfare of livestock: Sheep.* Ministry of Agriculture, Fisheries and Food, Department of Agriculture and Fisheries for Scotland, Welsh Office Agriculture Department. MAFF Publications, London.

Ministry of Agriculture, Fisheries and Food (1991). *Codes of recommendations for the welfare of livestock: Pigs.* Ministry of Agriculture, Fisheries and Food, Department of Agriculture and Fisheries for Scotland, Welsh Office Agriculture Department. MAFF Publications, London.

Roe, F. J. C. (ed.)(1983). *Microbiological standardisation of laboratory animals.* Ellis Horwood, Chichester.

Refinements in rabbit husbandry. (1993). Second report of the BVAAWF/FRAME/RSPCA/UFAW Joint Working Group on refinement. *Laboratory Animals,* **27**(4), 301–29.

Scott, W. N. (ed.)(1978). *The care and management of farm animals.* Baillière Tindall, London.

Whary, M., Peper, R., Borkowski, G., Lawrence, W., and Ferguson, F. (1993). The effect of group housing on the research use of the laboratory rabbit. *Laboratory Animals,* **27**(4), 330–41.

Chapter 7   Disease prevention and health monitoring in animals

Allen, A. M. (ed.)(1986). *Manual of microbiological monitoring of laboratory animals.* National Institute of Health, Betheseda, USA.

Coates, M. E. and Gustafssen, B. E. (ed.)(1984). The germfree animal in biomedical research. In *Laboratory animal handbooks, No. 9.* Laboratory Animals Ltd., London.

Festing, M. (1993). *International index of laboratory animals,* (6th edn.) Centre for Mechanism of Human Toxicity, University of Leicester.

Video. *Small animal necropsy dissection for animal health technicians* by B. C. Ward and J. D. Conroy (Mississippi State University). Royal Veterinary College, London.

Chapter 8   Biological data

Rispat, G., Slaoui, M., Weber, D., Salemink, P., Berthoux, C., and Shrivastava, R. (1993). Haematological and plasma biochemical values for healthy Yucatan micropigs. *Laboratory Animals,* **27**(4), 368–73.

Chapter 9   Handling of laboratory species

Anderson, R. S. and Edney, A. T. B. (ed.)(1991). *Practical animal handling.* Pergamon Press, Oxford.

Biological Council (1992). *Guidelines on the handling and training of laboratory animals.* Universities Federation for Animal Welfare, Potters Bar, Herts.

Video. *Restraint and handling of birds*. Allan White Memorial Video Library, St. Helens, Lancashire.

Chapter 10   Procedural data

BVA/FRAME/RSPCA/UFAW (1993). Joint working group on refinement. Removal of blood from laboratory mammals and birds. *Laboratory animals*, **27**, 1–22.

Clemons, D. J., Besch–Williford, C., Steffen, E. K., Riley, L. K., and Moore, D. H. (1992). Evaluation of a subcutaneously implanted chamber for antibody production in rabbits. *Laboratory Animal Science*, **42**, 307–11.

Hu, C., Cheang, A., Retnam, L., and Yap, E. H. (1993). A simple technique for blood collection in the pig. *Laboratory Animals*, **27**(4), 364–67.

McGuill, M. W. and Rowan, A. N. (1989). Refinement of monoclonal antibody production and animal wellbeing. *Institute of Laboratory Animal Resources News*, **31** (1), 7–10.

McGuill, M. W. and Rowan, A. N. (1989). Biological effects of blood loss: implications for sampling volumes and techniques. *Institute of Laboratory Animal Resources News*, **31**(4), 5–18.

McKellar, Q. A. (1989). Drug dosages for small mammals. *In Practice*, **11**, 57–61.

Nau, R. and Schunck, O. (1993). Cannulation of the lateral saphenous vein—a rapid method to gain access to the venous circulation in anaesthetized guinea-pigs. *Laboratory Animals*, **27**, 23–5.

National Institute of Health (1988). NIH intramural recommendations for research use of complete Freunds adjuvant. *Institute of Laboratory Animal Resources News*, **30**(2), 9.

Osebold, J. W. (1982). Mechanisms of action by immunological adjuvants. *Journal of the American Veterinary Medical Association*, **181**, 983–7.

Otto, G., Rosenblad, W. D., and Fox, J. G. (1993). Practical venepuncture techniques in the ferret. *Laboratory Animals*, **27**, 26–9.

Smith, B. L., Flåøyen A., and Embling, P. P. (1993). A simple gag for the intragastric dosing of guinea-pigs (*Cavia porcellus*). *Laboratory animals*, **27**(3), 286–88.

Stills, H. F. Jr. and Bailey, M. Q. (1991). The use of Freund's Complete adjuvant. *Laboratory Animals*, **20**, 25–30.

United Kingdom Coordinating Committee on Cancer Research (1988). Guidelines for the welfare of animals in experimental neoplasia. *Laboratory Animals*, **22**, 195–201.

Chapter 11   Recognition of pain and stress in laboratory animals

Association of Veterinary Teachers and Research Workers (1986). Guidelines for the recognition and assessment of pain in animals. *Veterinary Record*, **118**, 334–8.

Association of Veterinary Teachers and Research Workers (1989). *Guidelines for the recognition and assessment of pain in animals.* Universities Federation for Animal Welfare, Potters Bar, Herts.

Barclay, R. J. Herbert, W. J., and Poole, T. G. (1988). *The disturbance index: a behavioural method of assessing the severity of common laboratory procedures on rodents.* Universities Federation for Animal Welfare, Potters Bar, Herts.

Laboratory Animal Science Association (LASA) Working Party Report (1990). The assessment and control of the severity of scientific procedures on laboratory animals. *Laboratory Animals*, **24**, 97–130.

Manser, C. E. (1992). *The assessment of stress in laboratory animals.* Research Animals Department, RSPCA, Horsham.

Morton, D. B. and Griffiths, P. H. M. (1985). Guidelines on the recognition of pain, distress and discomfort in experimental animals and an hypothesis for assessment. *Veterinary Record*, **116**, 431–6.

National Research Council (1992). Recognition and alleviation of pain and distress in laboratory animals. National Academy Press, Washington, DC.

Stoskopf, M. A. (1983). The physiological effects of psychological stress. *Zoological Biology*, **2**, 179–90.

United Kingdom Coordinating Committee on Cancer Research (1988). Guidelines on the welfare of animals used in experimental neoplasia. *Laboratory Animals*, **22**, 195–201.

Wright, E. M., Marcella, K. L., and Woodson, J. F. (1985). Animal pain: evaluation and control. *Laboratory Animals*, **14**, 20–30.

Yoxall, A. T. (1978). Pain in small animals—its recognition and control. *Journal of Small Animal Practice*, **19**, 432–38.

Chapter 12   Anaesthesia of laboratory animals

Alexander, D. J. and Clark, G. C. (1980). A simple method of oral endotracheal intubation in rabbits. *Laboratory Animal Science*, **30**, 871–3.

Conlon, K. C., Corbally, M. T., Bading, J. R., and Brennan, M. F. (1990). Atraumatic endotracheal intubation in small rabbits. *Laboratory Animal Science*, **40**(2), 221–2.

Flecknell, P. A. (1987). *Laboratory animal anaesthesia.* Academic Press, London.

Green, C. H. (1979). Animal anaesthesia. In *Laboratory animal handbooks*, No. 8. Laboratory Animals Ltd, London.

Green, C. H., Knight, J., Precious, S., and Simpkin, S. (1981). Ketamine alone and combined with diazepam or xylazine in laboratory animals: a 10 year experience. *Laboratory Animals*, **15**(2), 163–70.

Hall, L. W. and Clarke, K. W. (1991). *Veterinary anaesthesia*, (9th edn). Ballière Tindall, London.

Jones, D. M. (1977). The sedation and anaesthesia of birds and reptiles. *Veterinary Record*, **101**, 340.

Lumb, W. V. and Jones, W. E. (1984). *Veterinary anaesthesia*. Lea & Febiger, Philadelphia.

Manser, C. E. (1992). The assessment of stress in laboratory animals. Research Animals Department, RSPCA, Horsham.

Samour, J. H., Jones, D. M., Knight, J. A., and Howlett, J. C. (1984). Comparative studies of the use of some injectable anaesthetic agents in birds. *Veterinary Record*, **115**, 6–11.

Short, C. E. (ed.) (1987). *Principles and practice of veterinary anaesthesia*, pp. 28–46. Williams & Wilkins, Baltimore.

Stuart, N. C. (1981). Anaesthetics in fishes. *Journal of Small Animal Practice*, **22**, 377–83.

Vickers, M. D., Schnieden, H., and Wood-Smith, F. G. (1984). *Drugs in anaesthetic practice*. Butterworth, Sevenoaks, Kent.

Chapter 13   Post-operative care and analgesia

Alexander, J. I. and Hill, R. G. (1987). *Post operative pain control*. Blackwell, Oxford.

Ciofalo, V. B., Latranyi, M. B., Patel, J. B., and Taber, R. I. (1977). Flunixin meglumine: a non-narcotic analgesic. *Journal of Pharmacology and Experimental Therapeutics*, **200**, 501–7.

Coles, B. H. (1984). Some considerations when nursing birds in veterinary premises. In *Journal of Small Animal Practice*, **25**, 275–88.

Dyson, D. H. (1990). Update on butorphanol tartrate: use in small animals. In *Canadian Veterinary Journal*, **31**(2), 120–1.

Flecknell, P. A. (1984). The relief of pain in laboratory animals. *Laboratory Animals*, **18**, 147–60.

Flecknell, P. A. (1991). Anaesthesia and post-operative care of small mammals. *In Practice*, **13**, 180–9.

Flecknell, P. A. and Liles, J. H. (1990). Assessment of the analgesic action of opioid agonist-antagonists in the rabbit. *Journal of the Association of Veterinary Anaesthetists*, **17**, 24–9.

Jenkins, W. L. (1987). Pharmacologic aspects of analgesic drugs in animals: an overview. *Journal of the American Veterinary Medical Association*, **191**, 1231–40.

Kitchell, R. L., Erikson, H. H., and Karstens, E. (1983). *Animal pain*. American Physiological Society, Bethesda.

Lees, P., May, S. A, and McKellar, Q. A. (1991). Pharmacology and therapeutics of non-steroidal anti-inflammatory drugs in the dog and cat: 1. General Pharmacology. *Journal of Small Animal Practice*, **32**, 183–93.

Liles, J. H. and Flecknell, P. A. (1992). The use of non–steroidal anti-inflammatory drugs for the relief of pain in laboratory rodents and rabbits: a review. *Laboratory Animals*, **26**, 241–55.

McKellar, Q. A. (1989). Drug dosages for small mammals. *In Practice*, **11**, 57–61.

McKellar, Q. A., May, S. A., and Lees, P. (1991). Pharmacology and therapeutics of non-steroidal anti-inflammatory drugs in the dog and cat. 2. Individual agents. *Journal of Small Animal Practice*, **32**(5), 225–35.

Michell, A. R., Bywater, R. J., Clarke, K. W., Hall, L. W., and Waterman, A. E. (1989). *Veterinary fluid therapy*. Blackwell, Oxford.

Otterness, I. G. and Gans, D. H. (1988). Non-steroidal anti-inflammatory drugs: an analysis of the relationship between laboratory animal and clinical doses, including species scaling. *Journal of Pharmaceutical Science*, **77**, 790–5.

Pekow, C. (1992). Buprenorphine jell-o recipe for rodent analgesia. *Synapse*, **25**(3), 35.

Rainsford, K. D. (1975). A synergistic interaction between aspirin or other non-steroidal anti-inflammatory drugs, and stress which produces severe gastric mucosal damage in rats and pigs. *Agents and Actions*, **5**, 553–8.

Shaw, K. *et al.* (1988). Analgesic and anaesthetic applications of butorphanol in veterinary practice. In *Proceedings, Western Veterinary Conference*, Las Vegas, Nevada. Vet Learning Systems Inc, Philadelphia.

Strub, K. M., Aeppll, L., and Müller, R. K. M. (1982). Pharmacological properties of carprofen. *European Journal of Rheumatology & Inflammation*, **5**, 478–87.

Taylor, P. M. (1985). Analgesia in the dog and the cat. *In Practice*, **7**, 5–13.

Universities Federation for Animal Welfare (1989). *Guidelines on the care of laboratory animals and their use for scientific purposes. 11. Pain, analgesia and anaesthesia*. Universities Federation for Animal Welfare, Potters Bar, Herts.

Chapters 14 and 15   Surgical and suturing techniques

Anderson, R. M. and Romfh, R. F. (1980). *Techniques in the use of surgical tools*. Appleton-Century-Crofts, New York.

Bradfield, J. F., Schachtman, T. R., McLaughlin, R. M. Steffen, E. K. (1992). Behavioural and physiologic effects of inapparent wound infection in rats. *Laboratory Animal Science*, **42**, 572–8.

Braun Melsungen AG. (1978). *Wound closure in the operating theatre*. Melsungen, Germany.

Cunliffe-Beamer, T. L. (1990). Surgical techniques. In *Guidelines for the wellbeing of rodents in research*, (ed. H. Guttman). Proceedings of a conference held at the Scientist Centre for Animal Welfare, North Carolina. SCAW, Bethesda MD

Dougherty, R. W. (1981). *Experimental surgery in farm animals.* Iowa State University Press, Ames, Iowa.

Harari, J. (1993). *Surgical complications and wound healing in the small animal practice.* W. B. Saunders, Philadelphia.

Hickman, J. and Walker, R. (1980). *An atlas of veterinary surgery,* (2nd edn). John Wright, Bristol.

Hurov, L. (1978). *Handbook on veterinary surgical instruments and glossary of surgical terms.* W. B. Saunders, Philadelphia.

Kirk, R. M. (1978). *Basic surgical techniques,* (2nd edn). Churchill Livingstone, Edinburgh.

Kirk, R. W. and Bistner, S. I. (1981). *Handbook of veterinary procedures and emergency treatment.* W. B. Saunders, Philadelphia.

Knecht, C. D., Allen, A. R., Williams, D. J., and Johnson, J. H. (1987). *Fundamental techniques in veterinary surgery.* W. B. Saunders, Philadelphia.

Lang, C. M. (1982). *Animal physiologic surgery,* (2nd edn). Springer, New York.

Lumley, J. S. P. *et al.* (1990). *Essentials of experimental surgery.* Butterworth, Sevenoaks, Kent.

Strachan, C. J. L. and Wise, R. (ed.) (1979). *Surgical sepsis.* Academic Press, London.

Swindle, M. M, and Adams R. J. (1988). (ed.) Experimental surgery and physiology: induced animal models of human disease. In *Experimental surgery and physiology.* Williams & Wilkins, Baltimore.

Turner, A. S. and McIlwraith, C. W. (1982). *Techniques in large animal surgery.* Lea & Febiger, Philadelphia.

Video. *Instrument care and sterilization* (1991). Royal Veterinary College, London.

Video. *Preparation and draping of the canine surgical patient.* Allan White Memorial Video Library, St. Helens, Lancashire.

Waynforth, H. B. and Flecknell, P. A. (1992). *Experimental and surgical technique in the rat,* (2nd edn). Academic Press, London.

Wind, G. G. and Rich, N. M. (1983). *Principles of surgical technique.* Urban & Schwarzenberg, Baltimore.

Glossary

Acidosis
: A reduction in the pH of the blood, commonly due to an increase in blood carbon dioxide level

Anterior
: An anatomical term meaning towards the front (usually head) end

Apnoea
: Respiratory arrest

Axenic
: Animals which are free from detectable microorganisms

Barbering
: The chewing and nibbling of whiskers and fur by one rodent to exert dominance over another

Cardiac tamponade
: Effusion of blood or fluid into the pericardium which prevents the heart from being able to fill effectively, thereby causing circulatory failure

Caudal
: An anatomical term meaning towards the tail

Coprophagy
: The eating of faeces carried out by rodents and lagomorphs to allow nutrients released during microbial digestion in the hindgut to be absorbed

Cranial
: An anatomical term meaning towards the head

Crepuscular
: Active at dawn and dusk

Cystocentesis
: A method of sampling urine by passing a needle through the abdominal wall into the lumen of the bladder (under aseptic conditions)

Dorsal
: An anatomical term meaning towards the back

Dyscrasia	Any disease causing abnormalities in the cellular constituents of blood
Dystocia	Difficulty giving birth
Embolism	The passage of a blood clot or other particulate matter in the bloodstream to a site distant from the site of origin, causing blockage of a vessel
Endotoxaemia	The presence of endotoxin (bacterial products) in the blood, which frequently results in shock, due to vasodilation
Exudate	The extravasation of fluid and/or cells from the blood into the tissues, any body cavity or the surface, usually due to inflammation
Gnotobiotic	Animals which harbour only known microorganisms
Haemolysis	The splitting or disintegration of red blood cells resulting in the release of haemoglobin
Hepatotoxin	A compound which is toxic to the liver
Hypercapnia	A high level of carbon dioxide in the blood
Hypnotic	As *Narcotic*. A drug which induces sleep
Hypoglycaemia	A low blood glucose level
Hypoxia	A low blood oxygen level
Infarction	Necrosis of part of an organ or tissue due to blockage of its arterial supply, by a thrombus or embolus in the end arteriole
Laparoscopy	The passage of an endoscope into the abdominal cavity, via a keyhole incision, for diagnosis or to perform surgical procedures
Laparotomy	A surgical procedure involving opening of the abdominal cavity
Lordosis	A position adopted by female animals in oestrus to facilitate copulation, with the back arched and the tail held to one side
Minute volume	Respiration rate $\times$ tidal volume

Narcotic	A substance which induces sleep
Nephrotoxin	A substance which is toxic to the kidney
Nystagmus	A repetitive movement of the eyeballs, consisting of a slow movement in one direction followed by a rapid return
Pheromone	Hormone-like substance secreted by one animal which causes behavioural, physiological, or endocrine changes in another
Phlebitis	Inflammation of a vein
Posterior	An anatomical term meaning towards the rear
Rostral	An anatomical term meaning towards the nose, used for describing structures on the head
Ruminal tympani	Distension of the forestomach (rumen) with gas
Sedative	A drug which produces a calming effect and drowsiness
Second intention healing	The healing of an open skin wound by proliferation of the dermis and epidermis to close the defect
Skin tenting	A phenomenon, usually caused by dehydration, in which a fold of skin remains raised for several seconds after being pinched
Specific pathogen-free	Animals free from particular (named) microorganisms
Stereotypy	The performance of repetitive, abnormal behaviour patterns, often indicating an inadequate environment
Teratogenesis	The production of abnormalities in a fetus, due to exposure of the dam to physical or chemical insult
Therapeutic index	The ratio between the lethal dose of a drug and its effective clinical dose
Thrombosis	The inappropriate formation of a blood clot

Tranquillizer	A drug which produces a calming effect without drowsiness
Transudate	The passage of fluid into the tissues or a body cavity, usually passively without inflammation
Urticaria	Nettle rash
Ventilation	A term used to describe forced breathing, by the application of positive pressure to the airways (intermittent positive pressure ventilation)
Ventral	An anatomical term meaning towards the belly
Xiphisternum	Cartilaginous process on the caudal end of the sternum

Index